AF566775

Polymer Science : A Text Book

First Published in 2008 by

Ane Books India

4821 Parwana Bhawan, 1st Floor
24 Ansari Road, Darya Ganj, New Delhi -110 002, India
Tel: +91 (011) 2327 6843-44, 2324 6385
Fax: +91 (011) 2327 6863
e-mail: anebooks@vsnl.com
Website: www.anebooks.com

For

CRC Press

Taylor & Francis Group
6000 Broken Sound Parkway, NW, Suite 300
Boca Raton, FL 33487 U.S.A.
Tel : 561 998 2541
Fax : 561 997 7249 or 561 998 2559
Web : www.taylorandfrancis.com

For distribution in rest of the world other than the Indian sub-continent

ISBN-10 : 1 42006 819 9
ISBN-13 : 978 1 42006 819 1

British Library Cataloguing in Publication Data
A catalogue record for this book is available from the British Library

Printed at Brijbasi Art Press, Noida

PREFACE

Polymer Chemistry is now taught as an independent subject in the Indian Universities and Engineering Institutes. With the opening of several thousands of manufacturing units for plastics, elastomeric materials and fibers, and establishment of various biomedical, defence and space research organizations, the demand for properly trained personnel in polymer science and technology has increased. Many chemical engineering and chemistry students wish to seek a career in the polymer field. This book caters to the needs of all such students with a background of organic chemistry, and thrusts on the initial approach of the subject.

The book is an outcome of the teaching and research experience of the authors in the realm of polymer chemistry for over 20 years. It is written with an aim to appreciate those fundamental principles of polymer science and engineering that are of practical relevance. We hope that the reader will obtain both a broad, unified introduction to the subject matter that will be of immediate practical value and a foundation for more advanced study. Though the arrangement of topics in the book is conventional but the explanations given are clear, the figures are carefully done and the important points are usefully repeated. We believe, the book will prove easier to read and learn from than the other short polymer science texts. For readers' convenience, each chapter begins with a table comprising the learning objectives and contents of the chapter. At the end of the book, some practice questions covering all the chapters are given.

Despite our careful preparation of this book, errors might have crept in. We would appreciate if they are pointed out, so that we can improvise upon them in the subsequent editions.

Authors

FOREWORD

Hermann Studadinger, a German Chemist, was the first to propound that a polymer molecule is made up of many small repeating units. For his discovery of chemical structure for natural rubber, he received the noble prize. Later on, synthetic polymeric materials began replacing natural polymeric materials owing to their light weight, mechanical strength and environmental stability. Nowadays, polymers have found practical applications in almost every walk of life including automotives, packaging, sanitary wares, cosmetics, pharmaceuticals, surgical sutures, conducting and semiconducting devices, information technology, wireless teleconmmunication etc. The last few decades have witnessed the tremendous growth in polymer industry.

One of the authors of this book, Dr. Anuradha Mishra has been my Ph.D. student and has been teaching polymer chemistry for many years. She is actively engaged in research in polymers for water treatment. This book will cater to the needs of undergraduate students of Chemistry and Chemical Engineering who wish to form a basic understanding of polymer science. Being the first edition, the authors would be looking forward to constructive and critical suggestions both in content and style, which can make the book more useful for education in polymer chemistry. Members of the academic community are, therefore requested to help in upgradation of this book.

G N Mathur
Visiting Professor
Department of Chemical Engineering
IIT, Kanpur
(Formerly, Director, DMSRDE, Kanpur)

CONTENTS

6. POLYMER CHARACTERIZATION 101

7. POLYMERIZATION PRACTICES 123

8. POLYMER PROCESSING 131

BASIC TERMS AND DEFINITIONS

Abrasion resistance

The ability of a polymer to withstand mechanical action, such as rubbing, scraping or erosion, which tends to remove material progressively from it's surface.

Age resistance

The resistance to deterioration by oxygen and ozone in the air, and by heat and light. A variety of anti-oxidants are available which provide protection to differing degrees against such deteriorating factors.

Alternating copolymer

A copolymer consisting of macromolecules comprising two species of monomeric units in alternating sequence.

Anionic polymerization

Anionic polymerization is that in which the kinetic-chain carriers are anions.

Block copolymer

In a block copolymer, adjacent blocks are constitutionally different, i.e., each of these blocks comprises constitutional units derived from different characteristic species of monomer or with different composition or sequence distribution of constitutional units.

Blow molding

Extrusion of a parison (hollow melt tube) which is forced to a mold caving by internal pressure and cooled down to form a hollow article. Used for the production of bottles, wide-mouth containers, petrol tanks etc.

Capillary rheometry

Actual capillary rheometry gives information about flow properties of polymer melts. Results are given as a viscosity function, i.e., viscosity as a function of shear rate or shear stress. Compared to MFR measurement the shear rate range is much wider and the rates in processing applications are reached. With many products, however, flow instabilities (melt fracture) at high rates limit the useful range. A capillary rheometer consists of a barrel, piston and die. The sample is put into the barrel, heated and pushed by the piston through the die. The piston is driven by hydraulic pressure which allows more flexible choice of rates compared to MFR measurements. In addition, longer dies can be used to avoid the errors coming from the entrance and exit pressure losses.

Catalyst

A substance which aids or promotes a chemical reaction without forming part of the final product. It enables the reaction to take place faster or at a lower temperature, and remains unchanged at the end of the reaction.

Chain polymerization

A chain reaction in which the growth of a polymer chain proceeds exclusively by reaction(s) between monomer(s) and reactive site(s) on the polymer chain such that the reactive site(s) are regenerated on the same polymer chain by the end of each growth step. A chain polymerization consists of initiation and propagation reactions, and may also include termination and chain transfer reactions. The adjective "chain" in "chain polymerization" denotes "chain reaction". In a chain polymerization, the average degree of polymerization remains constant with monomer conversion (e.g., in steady-state, radical polymerizations) or may increase with monomer conversion (e.g., in the formation of living polymers).

Chain transfer

A chemical reaction, usually occurring during chain polymerization, in which the activity of the kinetic-chain carrier is transferred from the growing macromolecule or oligomer molecule to another molecule or another part of the same molecule. Chain transfer to another part of the same molecule is often termed back-biting.

Comonomer

Comonomer is a monomer which is used to modify the polymer properties, e.g., to reduce the density of the polymer. Different types of copolymerisation processes are possible, leading to different structures and properties of the produced copolymer. In polyethylene, higher α-olefins like butene, hexene or octene are used as comonomers, in polypropylene, mostly ethylene and sometimes butene is used.

Compression set

The residual decrease in thickness of a test specimen measured 30 minutes or 24 hours after removal from a suitable loading device in which the specimen had been subjected for a definite time to compressive deformation under specified conditions of load application and temperature.

Conductive rubber

A rubber with a much lower electrical resistivity than normal. This is achieved by high loadings(e.g. 50phr) of one of the more highly conducting grades of carbon black, e.g. super conduction furnace or acetylene black. A typical resistivity is about 1015¥Øcm compared with about 1015¥Øcm in a normal rubber.

Copolymer

Two monomers polymerized together to form a polymer.

Creep tests

1. Solid-state creep tests are used to measure long term behavior of mechanically loadedmaterial. Creep is time-dependent strain resulting from constant loading. Tensile loading is most common here but flexural loading can be used as well, testing time is typically1000 h. Several loading levels are used to get an overview of creep behavior of the material.
2. Melt creep tests are a type of rheological measurement for the determination of the creep compliance of a polymer melt; they are carried out in the region of linear viscoelasticity.

Crosslink

A constitutional unit connecting two parts of a macromolecule that were earlier separate molecules. Note: a network may be thought to consist of many "primary chains" that are interconnected by a number of crosslinks. In the vast majority of cases, the crosslink is a covalent bond but the term is also used to describe sites of weaker chemical interactions, portions of crystallites, and even physical entanglements.

Crystallinity

Also designated *i*°degree of crystallinity¡±, it gives the relative fraction of a solid polymer having crystalline regularity. The crystallinity can be determined with various measuring techniques like DSC, density measurement or WAXS, which has the additional advantage that the relative amounts of the different crystal modifications can be determined as well. All crystallinity values are relative and the applied reference values should be quoted.

Density

An important parameter for polyethylene, affecting most end product physical properties like stiffness, impact strength and optical properties. The density is normally given in g/cm^3, although the correct SI unit is kg/m^3. Two basic measuring methods are used: (a) Density gradient column is the basic method for density measurement. In it a density gradient is prepared by mixing two liquids of different densities so that density increases uniformly from top to bottom. Small test pieces are immersed into this density gradient, and that the density of the polymer is determined based on the equilibrium position of the pieces and calibration floats. (b) Pyknometer measurements rely on calibrated volumes and the density difference to a reference fluid; compression-moulded samples are normally used here.

Differential Scanning Calorimetry (DSC)

In DSC the sample and a reference material are heated and cooled in a defined manner. DSC provides a rapid method for the determination of the temperature profile of a polymeric material. This includes measurement of heat capacity, specific heat, melting, crystallization and glass transition temperatures, heat (enthalpy) of fusion, crystallization, study of thermal stability or reaction kinetics and in certain cases identification of the polymer or polymer mixture.

Double-strand polymer

Ladder polymer: a polymer, the macromolecules of which are double-strand chains.

Durometer

An instrument for determining the hardness of a polymer by measuring it's resistance to the penetration (without puncturing) of a blunt indented point impressed on the rubber surface against the action of a spring. A hand and a special scale indicate the resistance to penetration or "hardness". The scale of the Shore hardness testers reads from 0 to 100, 0 being very soft and 100 being very hard. There are several Shore hardness gauges available and the proper durometer depends on the hardness of the surface being measured (Shore A, D and 00).

Dynamic Mechanical Thermal Analysis (DMTA)

Many polymeric materials exhibit time-dependent, reversible viscoelastic properties in deformation. Dynamic mechanical tests are characterized by an application of a small stress in sinusoidal fashion and a continuous scan of resonant frequency of vibration and damping vs. temperature is produced. This is transformed into dynamic storage modulus, dynamic loss modulus and damping factor by the instrument software.

Dynamic rheometry

Dynamic rheometry gives information both about flow and elastic properties of polymer melts. In a dynamic measurement the sample is put between two round plates or between a cone and a plate in the oven. The system is heated to a desired temperature and a sinusoidal deformation at different frequencies is applied. As a result, we get storage and loss modulus as a function of frequency. Storage modulus is connected with elastic energy, while loss modulus corresponds to viscose energy. From the above parameters we can calculate complex viscosity as a function of frequency.

With most of the polyolefin products this is the same as the conventional viscosity function, which is viscosity as a function of shear rate. The dynamic measurements are limited to relatively low shear rates. On the other hand, a benefit over capillary rheometry is that we also get reliable information about elasticity. In addition, the obtained properties are much more sensitive for small differences in polymer structure which play an important role in processing and end-use properties.

Elastomer

A polymer with the properties of rubber. Polymers that can be formulated as elastomers are polyurethane, butyl rubber, silicones and specially treated ethylene-propylene copolymers.

End-group

A constitutional unit with only one attachment to a chain.

Expanded polymer

Alternate name for cellular polymer, but the term is often restricted to those cases where the material has been produced by allowing a gas to expand within a polymer melt and then cooling the melt, thus trapping the gas bubbles in the solid polymer formed. The gas is produced either by the injection under pressure or by the chemical decomposition, usually induced by the high temperatures of melt processing of a blowing agent.

Extruder

Equipment for melting, pressurising and homogenising plastics by means of a rotating screw. Different configurations are possible; the most simple one being a single screw extruder mainly used for the conversion processes like cast film extrusion, fibre spinning, film blowing or pipe and profile extrusion. For more demanding applications like compounding of polymer powder before pelletization or the mixing of polymers, twin screw extruders are used where the two normally intermeshing screws can be moved in a corotating (same direction) or counterrotating (opposite direction) fashion. These machines normally contain different mixing elements, most important being kneading blocks for the local creation of extensional flow.

FTIR spectroscopy

Infrared radiation induces vibrations of molecules and molecular segments. Absorption of radiation occurs only if the vibration results in the change of the molecular dipole moment. The strength of the vibrating bonds and their dipole moment affect the absorption band intensities and positions which makes FTIR spectroscopy a very useful method in qualitative polymer analysis. Also several fast quantitative FTIR methods have been developed for the structural characterisation of polyolefins. Polymer crystallinity and orientation implies a restriction in the movement of some molecules and consequent improvement in polymer mechanical properties. This restriction in the movements also reflects in the FTIR spectrum and thus the amorphous and crystalline polymer phases can be identified.

Flammability

The characteristics of a material that pertain to it's relative ease of ignition and relative ability to sustain combustion.

Flexural test

Like tensile test, flexural test is also used to characterize strength and stiffness of the material. Flexural properties are obtained by placing a specimen horizontally on two supports. A load is applied at a specified rate in the center, normal to specimen longitudinal axis. Measured force - deformation data is used to calculate flexural modulus and flexural strength. Flexural tests can be used if material is too brittle to withstand compressive forces of fixing in tensile grips.

Flory-Huggins theory

A thermodynamic theory of polymer solutions, first formulated independently by Flory and by Huggins, in which the thermodynamic quantities of the solution are derived from a simple concept of combinational entropy of mixing and a reduced Gibbs-energy parameter, the 'X parameter'. The X parameter is a numerical parameter employed in the Flory-Huggins theory, which accounts in the main for the contribution of the non-combinational entropy of mixing and for the enthalpy of mixing.

Gel Permeation Chromatography (GPC)

GPC is used to separate dissolved polymer molecules according to their coil diameter in solution. The chromatograms can be converted to molar mass distributions, which is an important parameter because it has a significant effect on the physical, mechanical and rheological properties of a polymer. It is also possible to calculate the average molar masses, M_n, M_v, M_w and M_z, and when on-line viscometric detection is used, to study long chain branching (LCB) and its distribution.

Graft polymer

A substance composed of graft macromolecules

Hardness

Resistance of a (polymer) surface to deformation. The different hardness measures applied for characterising polymers are:

(a) Shore hardness (two scales, A for softer and D for harder materials)
(b) Ball indentation hardness (also useable on profiled surfaces because of bigger measuring device).

Heat Deflection Temperature (HDT)

Heat deflection temperature (HDT) indicates how loaded material deforms at higher temperatures. Test bars are placed in a heating bath, resting horizontally on two supports. A constant load is applied in the center of the specimen and the bath temperature is raised at a constant rate. The temperature of the bath at which the flexural deflection of the loading point has reached a predefined level is the heat deflection temperature of the material.

High Density Polyethylene (HDPE)

Coming from a low pressure process based on special catalysts (Ti- or Cr-based), HDPE has a linear polymeric chain, combined with a higher density and melting point. To fit special applications, the molar mass distribution can be adjusted in a wide range by means of catalysts or polymerization technology. New applications are opened for HDPE as well as other low pressure PEs by applying metallocene (single-site) catalysts, which allow the production of very narrow molar mass distributions and the better distribution of comonomers.

Homopolymers

Homopolymers (HOMO) consist only of one monomer units; their properties can be regulated via molar mass distribution and the degree of isotacticity (stereoregular structure).

Impact strength

Impact strength characterizes material behaviour under high speed loading. Pendulum and falling weight type testers are applied here. Specimen can be either plaques, notched or unnotched bars or parts of finished products. There are several impact methods like Charpy impact test, instrumented puncture test and tensile impact test. Generally impact test shows the energy which is needed to break or puncture specimen under specified conditions. In some tests it is possible to accurately record force - deformation transient curves and calculate specific parameters describing material behavior under impact loading.

Injection molding

A repetitive process in which plastic is melted and injected into a mold cavity where the article is cooled down. After cooling, the mold opens and the article is ejected. The melt is prepared in a screw setup similar to an extruder; the screw acts in melting and homogenizing the polymer while slowly retracted to build up the melt reservoir necessary for the injection step. Then, the screw is used as a plunger in a forward movement to inject the melt through the runner, a manifold in case of a multiple mold and the gate into the mold. In the injection molding process, filling of the mold and solidification takes place partially at the same time, resulting in a very complex morphology in case of semicrystalline polymers.

Ionic polymerization

A chain polymerization in which the kinetic-chain carriers are ions. Usually, the growing chain ends are ions.

Liquid rubber

A liquid, usually a viscous low molecular weight polymer (typical molecular weight being a few thousand), which may be readily crosslinked to a product similar in behaviour to that of a normal rubber vulcanizate. The use of a liquid rubber offers the considerable technological advantage over conventional rubbers that expensive compounding and press vulcanization procedures are not required. Some polyurethane elastomers, silicone rubbers and polysulphide rubbers may be processed as liquid rubbers. An early liquid rubber was depolymerised natural rubber and the term originally applied specifically to this material. More recently liquid rubbers based on polybutadiene (of molecular weight about 10000), which may be crossliked through functional groups on the chain ends(telechelic polymer), have been developed, e.g., carboxy-terminated polybutadiene and hydroxy-terminated polybutadiene and similar nitrile and styrene-butadiene copolymers.

Liquid crystalline side chain polymer

A liquid crystalline polymer in which the mesogenic groups are part of the groups attached to the main chain. Ordering of the mesogens, and hence liquid crystalline behavior, is enhanced by distancing them from the main chain with flexible groups.

Living polymerization

A chain polymerization in which the concentration of kinetic-chain carriers, under the appropriate conditions for synthesis, remains constant for a period many times longer than the duration of the synthetic procedure. Often, the absence of chain transfer is implied in the term "living polymerization".

Lower Critical Solution Temperature (LCST)

The critical solution temperature below which a binary mixture exists as a homogeneous single phase solution at all compositions. Thus when such a mixture existing below the LCST is heated, phase separation occurs. Such behavior is more unusual than upper critical solution temperature behavior and is not accounted for by the simple theories of solution, e.g., the Flory-Huggins theory for polymer solutions. However, it does occur for several polymer/solvent pairs and not at all infrequently for polymer/polymer pairs, i.e., polymer blends. This behavior implies an increase in x(the Flory-Huggins interaction parameter) with temperature, and can be accounted for by the corresponding states theory, as being due to dissimilar thermal expansion coefficients between polymer and solvent. At high temperatures, the more highly expanded solvent must fit into a denser medium and mixing must involve a negative volume change and negative entropy of mixing.

Macromolecule

A molecule of high relative molecular mass, the structure of which essentially comprises the multiple repetition of a number of constitutional units

Melt Flow Rate (MFR)

Melt mass flow rate or melt flow rate (MFR) gives an indication of melt processing properties of a polymer. Shear rate range is, however, narrow and much lower than in the most of processing applications. It gives thus very limited information about melt flow behavior. The method is a kind of capillary measurement. The equipment consists of a barrel, a piston with a place for loading masses and a die. The sample is charged into the barrel and heated up to a specified temperature. After that the piston is loaded extruding the polymer through the die. Melt flow rate (MFR) is defined as a mass flown through the die in 10 minutes. The weights mostly used with polyolefins are 2.16 kg, 5 kg, 10 kg and 21.6 kg. In addition to the evaluation of flow properties MFR serves as a rough estimation of molecular weight. Usually Flow Rate Ratio (FRR) is also determined. This is defined as a ratio of two MFR values obtained with two different weights. FRR can be used as a rough estimation of molecular weight distribution.

Metallocene

A catalyst which contains a metallocene compound of a transition metal. Often used as a synonym for single site.

Molecular Weight Distribution (MWD)

The relation between the number of molecules in a polymer and their individual chain length. It is often called molar mass distribution (MMD). Technical polymers are always polydisperse, as a result of the nature of the catalyst and the polymerization conditions. The MWD is often characterised by size exclusion chromatography (SEC), but it can also be calculated approximately from rheological data for purely linear polymers.

Monomer

A molecule capable of reacting with like or unlike molecules to form a polymer.

Morphology

Morphology is the internal or external structure of a solid substance. Important sub-definitions include
(a) Crystal morphology of a polymer film or part (b) Phase morphology of heterophasic polymer systems (blend, copolymer) (c) Powder morphology of a polymer powder from the polymerization reactor.

Nucleating agents

Substances with the ability to act as substrates for crystal formation by epitaxy, thus increasing the nucleation density. Improve crystallization behavior in processing, i.e., cycle time or line speed, as well as crystallinity and morphology in final product, improving optical and mechanical properties.

Olefin

An unsaturated simple hydrocarbon with one double bond per molecule, such as ethylene or butylene. Olefins are chained into polymers such as polyethylene.

Olefin fiber

Generic name for fiber composed of a polymer with at least 85 wt per cent of ethylene, propylene or other olefin units, excepting amorphous rubbery polymers. Examples are Courlene, Fibralon, Herculon and Vectra.

Oleoresin

The initial exudation from certain trees when the bark is wounded. An oleoresin consists of a resinous component together with an essential oil. The latter usually evaporates, leaving the hard resin. Best studied is pine oleoresin, from which the resin—rosin and the essential oil turpentine are obtained.

Oligomer

A substance composed of oligomer molecules.

Permeability

Material constant characterising gas penetration, like water vapour, oxygen or carbon dioxide.

Peroxide

Peroxides are a group of compounds that contain the O^{2-} ion. It is used to initiate the polymerization for high-pressure free radical polymerization processes.

Polyaddition

A polymerization in which the growth of a polymer chain proceeds by addition reactions between molecules of all degrees of polymerization, not accompanied by the formation of low-molar-mass by-product(s).

Polycondensation

A polymerization in which the growth of a polymer chain proceeds by condensation reactions between molecules of all degrees of polymerization accompanied by the formation of low-molar-mass by-products(s). The growth steps are expressed by P_x+P_y → P_x+y + L {x} element of {1, 2, ... ¡Ä}; {y} element of{1, 2, ...¡Ä} where P_x and P_y denote chains of degree of polymerization x and y, respectively, and L a low-molar-mass by-product. In a polycondensation where the total amounts of the monomers are present from the beginning of the polymerization, the average degree of polymerization increases with conversation of reactive groups.

Polymer

A substance consisting of molecules characterized by the repetition (neglecting ends, branch junctions and other minor irregularities) of one or more types of monomeric units.

Polyolefin

This is the collective name given to those polymers that are made from the lower olefins: ethylene, propylene, butylene and isoprene. The polyolefins are thermoplastic polymers.

Prepolymerization

A process where polymerization is started to increase the activity of the catalyst getting over the induction time of the catalyst and also not to overheat the catalyst particles to trigger sudden onset of polymerization. Prepolymerization also helps in ensuring even more polymerization on the catalyst particles reducing the probability of creating a high fraction of fines in the later reaction steps.

Processability

A concept to describe the behavior of the polymer in a converting process, such as blow molding machine and film line. A good processability means a stable behavior in the process and a high processing speed.

Pseudo-copolymer

An irregular polymer, the molecules of which are derived from one species of monomer but for which one finds it convenient to use copolymer terms. Where appropriate, adjectives specifying the types of "copolymer" may be applied to "pseudo-copolymer". The term statistical pseudo-copolymer, for instance, may be used to describe an irregular polymer whose molecules have the sequential distribution of configurational units obeys known statistical laws.

Random copolymers

Random copolymers are produced by statistical insertion of units of ethylene or higher—olefins (mostly butene, also in combination with ethyle to give terpolymers). Through this, the glass transition temperature and melting point are lower than that in homopolymers, allowing better sealability and application at temperatures below zero degree.

Reclaimed rubber

The product resulting from the treatment of vulcanized scrap rubber in various operations, such as grinding, defibering and devulcanizing by heating with aqueous alkali or with a metallic chloride solution and reclaiming agents, and refining, whereby the fabric is destroyed and the rubber compound is recovered and made suitable for use in the manufacture of rubber goods; the fiber may also be removed mechanically, followed by devulcanization of the rubber.

Resin

Any of the class of solid or semi-solid organic products of natural or synthetic origin, generally of high molecular weight with no definite melting point. Most resins are polymers.

Ring-opening polymerization

A polymerization in which a cyclic monomer yields a monomeric unit which is acyclic or contains fewer cycles than the monomer. If the monomer is polycyclic, opening of one ring is sufficient to classify the reaction as ring-opening polymerization.

Scanning Electron Microscopy (SEM)

In Scanning Electron Microscopy (SEM) a focused beam of electrons scan across the surface of the sample. It is detected synchronously to the position of the scanning electron beam and recorded in an image storage device. Differences in the topography of the sample give rise to the differences in the intensities of the signals formed by the secondary electrons in the detector resulting in a three dimensional image. Scanning electron microscopy is suited to most types of surface morphology studies. Rough topographic features, void content and particle agglomerations are easily revealed as well as the compositional differences within material. Energy Dispersive X-ray Spectroscopy (EDS) or Wavelength Dispersive X-Ray Spectroscopy (WDS) in conjunction with SEM allows elemental analyses to be performed directly over the surface of the sample.

Shear thinning

A behavior where the viscosity decreases when the shear stress increases. Typically broadening of the molecular weight distribution increases shear thinning.

Shore hardness

Resistance of a (polymer) surface to deformation. The different hardness measures applied for characterising polymers are:

(a) Shore hardness (two scales, A for softer and D for harder materials)
(b) Ball indentation hardness (also useable on profiled surfaces because of bigger measuring device).

Solution Viscometry

In Solution Viscometry the rate at which a dilute polymer solution flows through a capillary is measured. Temperature and concentration of the polymer are carefully controlled. The parameter most often determined by dilute solution viscometry is intrinsic viscosity. Intrinsic viscosity is related to molar mass through a semi-empirical relationship called Mark-Houwink's equation.

Sponge rubber

Vulcanized rubber having a porous or cellular structure like that of a sponge, made by incorporating gasifying substances or blowing agents such as sodium bicarbonate and ammonium carbonate in the mixing, and then vulcanizing in a mold cavity larger than the piece of rubber to be vulcanized. Gas is generated from the dispersed particles of the blowing agent, producing the porous structure and inflating the rubber to the size of the mold cavity.

Star macromolecule

A macromolecule containing a constitutional unit from which more than two chains (arms) emanate. A star macromolecule with *n* linear chains (arms) attached to the central unit is termed an *n*-star, e.g., five-star.

Swelling

The property of a raw or vulcanized rubber of absorbing organic liquid, such as benzene and gasoline, and swelling to many times it's original volume; the property is also shown by other colloids in contact with the various liquids; in a general sense, it may be any increase in volume in a solid substance caused by the absorption of a liquid.

Syndiotactic macromolecule

A macromolecule comprising alternating enantiomeric configurational base units. ***Note:*** In a syndiotactic macromolecule, the configurational repeating unit consists of two configurational base units that are enantiomeric.

Tacticity

The orderliness of the succession of configurational repeating units in the main chain of a regular macromolecule (or oligomer or block).

Tear resistance

The force required to tear completely across a specifically nicked rubber test specimen, or right angle test specimen, by elongating at a specific rate.

Tear strength

Film must have certain resistance to tear loading. Tear strength indicates the average force required to propagate tearing through a specified length of the film sample. This test is suitable for investigation of orientation balance of the film. Normally tear strength is measured in two directions, parallel and transverse to the extrusion direction of the film.

Telomer

A substance composed of macromolecules or oligomer molecules having few, usually terminal, reactive functional groups enabling, under appropriate conditions, the formation of larger macromolecules.

Thermoforming

The process of heating a thermoplastic sheet to a working temperature and then forming it into a finished shape by means of heat or pressure.

Thermogravimetry (TG, TGA)

Thermogravimetry (TG, TGA) can be used to measure any reaction involving mass change. TG analysis is often complementary test to DSC, the weight loss curve giving a picture of the make-up of the material by showing the temperatures at which individual components volatilize.

Thermomechanical Analysis (TMA)

Thermomechanical Analysis (TMA) is a technique for monitoring changes of physical dimensions (expansion, shrinkage, penetration) while the temperature of the sample is changed.

Vicat softening temperature

Vicat softening temperature shows heat softening characteristics of thermoplastics. Flat specimen are placed in a temperature regulated heating bath, a needle type, loaded penetrator is set on the specimen surface and bath temperature is raised at a constant rate. The temperature of the bath at which the penetration of the needle has reached a predefined level is the Vicat softening temperature of the material.

Vitrification

Point at which a system becomes a glassy solid, and its molecules become immobilized, except for vibrations. At vitrification, the reaction rate (kinetics) drops off quickly.

Vulcanization

An irreversable process during which a rubber compound, through a change in it's chemical structure (for example, cross-linking), becomes less plastic and more resistant to swelling by organic liquids and elastic properties are conferred, improved or extended over a greater range of temperature.

Ziegler-Natta catalyst

Ziegler-Natta catalyst produces polymer having relatively narrow MWD. Aluminum alkyl cocatalyst is required.

1

INTRODUCTION

Learning Objectives

After studying this chapter, you will know:

- What are polymers?
- What various terms are used in polymer science?
- What are the bases for classification of polymers?
- What types of isomerism do exist in polymer molecules?
- What are copolymers?
- What types of bonding exist in polymer molecules?

Chapter Contents

• Polymer • Monomer • Degree of polymerizations • Oligomers • High polymers • Line, branched and crosslinked polymers • Thermosets and thermoplastics • Isomerism in polymer molecules • Addition and condensation polymers • Bonding in polymers • Flow chart of polymeric materials

1.1 DEFINING POLYMERS

What is a Polymer?

Poly comes from the Greek word for "many" and mer comes from the Greek word for "parts." A polymer is a large molecule that comprises repeating structural units joined by the covalent bonds, or a polymer is a group formed by repeated linking of small molecules. This group of many small molecules behaves as a single molecule. The small molecule or repeating unit or the building block or structural unit of the polymer is called monomer.

Monomer

To be a monomer, a chemical substance should have a functionality of at least two. The functionality of a compound depends on the number of reactive sites it has. Apart from compounds containing functional group, some other compounds having double or triple bonds in the molecule can also be polyfunctional, e.g., ethylene, etyrene and Terephthalic acid.

Ethylene can take on two hydrogens therefore it is bifunctional. So the definition of functionality is the number of hydrogen or halogen which can react with a molecule having double or triple bond or the number of replaceable hydrogen atoms in a molecule. For polyethylene, the building block or structural unit is

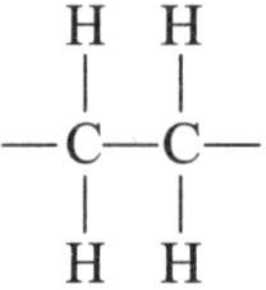

Polymerization

It is a chemical reaction in which two or more than two molecules of one or more than one substances combine together to form a molecule of high molecular weight (size 10^{-4} to 10^{-7} cm), e.g., $n\,CH_2 = CH_2 \rightarrow [CH_2\text{–}CH_2]_n$.

Introductory organic chemistry texts talk about a **homologous series** that contains methane, ethane, propane, ..., decane, ..., and so on.

This homologous series can be thought of as two methyl groups, $-CH_3$, tied together by many methylene groups, $-CH_2-$. As one extends this series, polyethylene is created.

$$CH_3-CH_2-CH_2-CH_2- \ldots -CH_2-CH_2-CH_3$$

or

$$CH_3-[-CH_2-]_n - CH_3 \quad \text{(a shorthand method)}$$

or finally

$$[-CH_2-]_n$$

n is a subscript to count the number of repeat units. This last shorthand method requires that one remembers to add a hydrogen to each end group (CH_3 on the end rather than CH_2).

Degree of Polymerizations

Number of monomer or repeating unit (n) in the polymer chain is called degree of polymerization, e.g., $[CH_2\text{–}CH_2]_n$.

Degree of polymerization (DP) is used to calculate the average molecular weight of polymer.

Average molecular weight of polymer = DP × Weight of repeating unit

Oligomers

Polymers which have comparatively low degree of polymerization are called oligomers, e.g., liquid polymers of molecular weight of the order of 2,000 or so, e.g., polyurethane.

High Polymers

Polymers with a high DP are called high polymers. The molecular weight of high polymers may vary from 10^4 to several millions, e.g., polystyrene.

1.2 CLASSIFICATION OF POLYMERS

There are several bases on which polymers can be classified.

1.2.1 On the Basis of Their Occurrence

Although on the basis of their occurrence polymers can be divided into two broad categories, natural and synthetic but a third category, semi-synthetic, also exist in between these two.

Natural Polymers Polymer molecules that are found in nature are called natural polymers, e.g., carbohydrates, proteins, quartz etc.

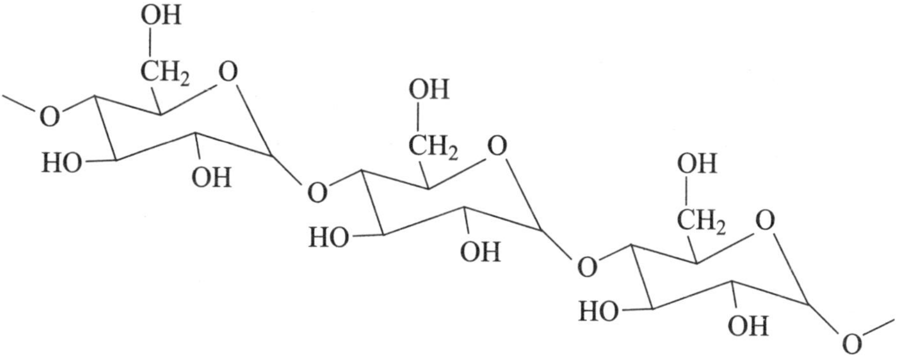

Starch molecule

Synthetic Polymers Man made polymers are called synthetic polymers, e.g., polystyrene, polymethylemethacrylate, nylons, synthetic rubbers etc.

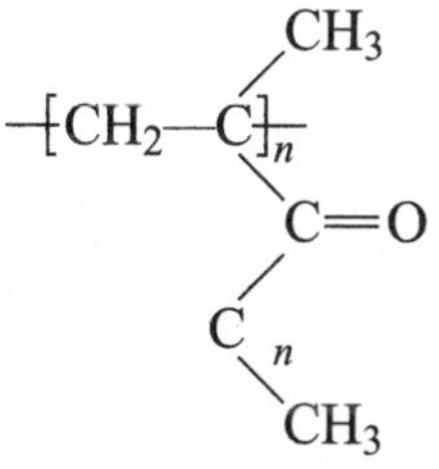

Polymethylmethacrylate

Semisynthetic Polymers These polymers either have back bone of any natural polymer and the pendant groups are substituted to make it perform better properties or have one of its components as natural polymer and other as synthetic, e.g., grafted copolymers of starch and polyacrylamides, silicones etc.

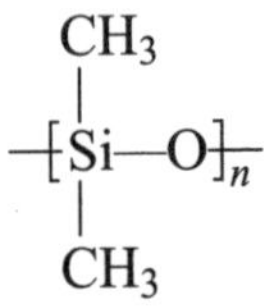

$$\left[\begin{array}{c} CH_3 \\ | \\ Si-O \\ | \\ CH_3 \end{array}\right]_n$$

Silicone

1.2.2 On the Basis of Types of Monomers

On this basis, polymers can be classified into two classes.

Homopolymer A polymer containing a single type of repeat unit is called a homopolymer, e.g., polystyrene.

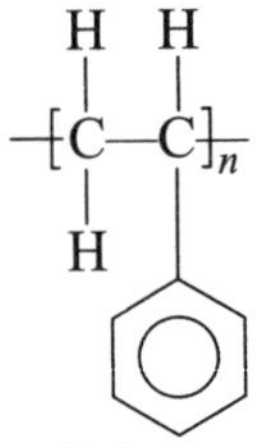

$$\left[CH_2-CH(C_6H_5)\right]_n$$

Polystyrene

Copolymer If a polymer is made up of two different monomers then it is called copolymer, e.g., styrene butadiene (SBS) rubber and Sty-co-AN.

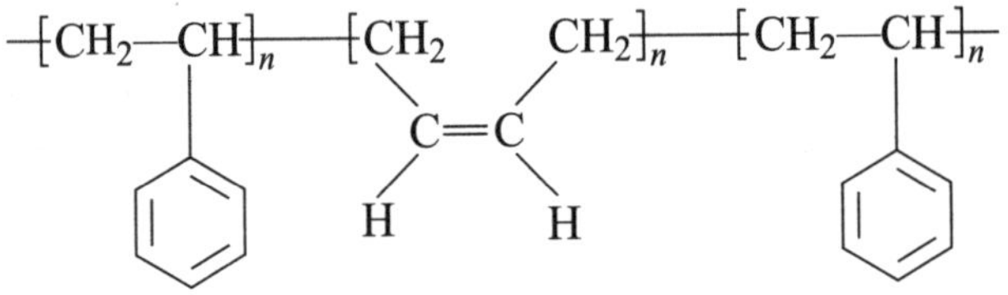

$$\left[CH_2-CH(C_6H_5)\right]_n-\left[CH_2-CH=CH-CH_2\right]_n-\left[CH_2-CH(C_6H_5)\right]_n$$

SBS rubber

1.2.3 On the Basis of Polymerization Process

Carothers in 1929 classified polymers into two types on the basis of mechanism of polymerization reaction:

(i) addition polymers
(ii) condensation polymers

But there were some polymers that didn't fit into either of the two strict categories proposed above. Therefore, these terms were modified by H.F. Mark on the basis of kinetics of polymerization (1950) as

(i) chain polymerization
(ii) step polymerization

On this basis, polymers can be classified into two classes—

Chain Growth Polymers Chain polymers are the products of self addition reaction of monomer molecules through a chain reaction. No by-product is formed in this reaction, the molecular weight of the polymer is the exact multiple of the molecular weight of monomer, e.g., all the vinyl monomers.

$$\underset{\text{styrene}}{H_2C=CH(C_6H_5)} \xrightarrow[\text{vinyl polymerization}]{\text{free radical}} \underset{\text{polystyrene}}{-\!\!\left[CH_2-CH(C_6H_5)\right]\!\!-_n}$$

Step Growth Polymers A step growth polymer is the product of a reaction that involves a random reaction of two molecules that may be any combination of a monomer, oligomer or a long chain molecule. It proceeds by step-wise fashion. Some of the examples of step growth polymers are nylons, polyurethanes, polyesters etc.

$$Na^+O^-\text{—}C_6H_4\text{—}C(CH_3)_2\text{—}C_6H_4\text{—}O^-Na^+ \; + \; Cl\text{—}C(=O)\text{—}Cl$$

$$\downarrow$$

$$-\!\!\left[O\text{—}C(=O)\text{—}O\text{—}C_6H_4\text{—}C(CH_3)_2\text{—}C_6H_4\right]\!\!-_n + (n-1)\,NaCl$$

$$\downarrow$$

$$-\!\!\left[O\text{—}C(=O)\text{—}O\text{—}C_6H_4\text{—}C(CH_3)_2\text{—}C_6H_4\right]\!\!-_n$$

1.2.4 On the Basis of Line Structure

The monomeric units in a polymer may be present in linear, branched, or crosslinked fashion.

Linear Polymer A straight chain species, i.e., the units are connected to each other in a chain arrangement. Linear polymers may be represented by a chain with two ends, e.g., high density polyethylene.

Branched The main chains of polymer molecules have small branches of the same or different molecules. Branched polymers have side chains, or branches, of significant length which are bonded to the main chain at branch points, and are characterized in terms of the number and size of the branches. Polymers with well-controlled modes of branching, e.g., comblike polymers, star polymers, dendrimers, low density polyethylene, graft copolymers etc.

Branched polymer

Sometimes the ends of several polymer chains are joined together at a common center. Polymers like this are called star polymers.

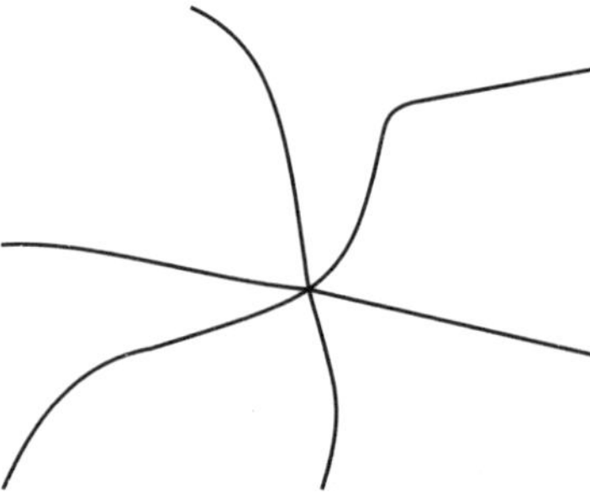

Star polymer

Sometimes there is no backbone chain at all and a polymer is built in such a way that branches just perpetuate i.e. more and more branches keep growing out. These are called dendrimers, from the ancient Greek word for "tree". Dendrimers have unusual shapes that make pretty pictures, and they also have really unusual properties.

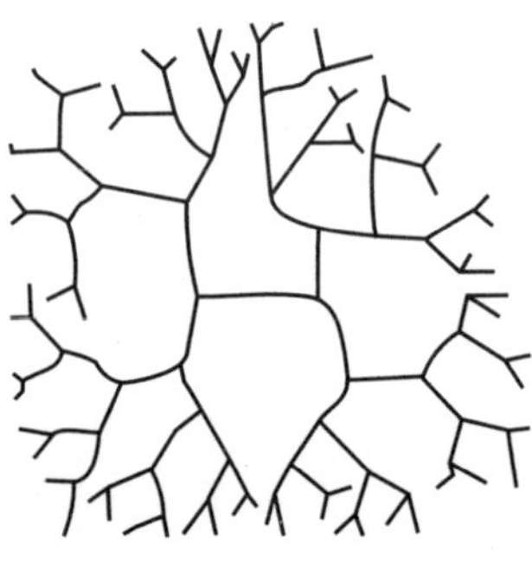

Dendrimer

Crosslinked Those polymers in which two linear chains are joined together by covalent bonds i.e. the chains are crosslinked. Network polymers have three-dimensional structures in which each chain is connected to all others, and are characterized by their crosslinking density or degree of crosslinking (number of junction points per unit volume).

The crosslinks tie all the polymer molecules together. Because crosslinked polymers are tied together they cannot flow past each other or around each other when heated and hence they do not melt. They get charred instead. Also, because all the polymer molecules are tied together, they aren't easily broken apart from each other. This is why the crosslinked polymers do not get brittle when they get cold.

Cross linked materials do not dissolve in solvents, because all the polymer chains are covalently tied together. But they can absorb solvents. A piece of a crosslinked material that has absorbed a lot of solvent is called a gel. Crosslinked polymers can also be coatings, adhesives, and electronic parts. e.g., vulcanized rubbers Novolac, melamine-formaldehyde resins etc.

What happens in rubbers during vulcanization? What does the sulfur do to the rubber? What it does is that it forms bridges which tie all the polymer chains in the rubber together. These are called crosslinks. As shown in the picture below, bridges made by short chains of sulfur atoms tie one chain of polyisoprene (natural rubber) to another, until all the chains are joined into one giant supermolecule.

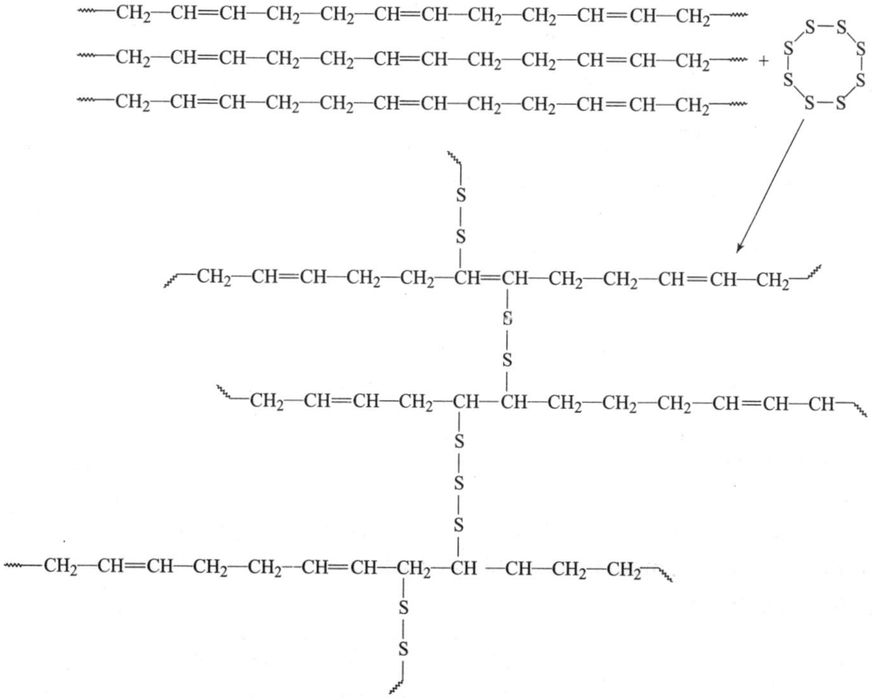

$\sim CH_2-CH=CH-CH_2-CH_2-CH=CH-CH_2-CH_2-CH=CH-CH_2\sim$

$\sim CH_2-CH=CH-CH_2-CH_2-CH=CH-CH_2-CH_2-CH=CH-CH_2\sim$ + S_8 (ring)

$\sim CH_2-CH=CH-CH_2-CH_2-CH=CH-CH_2-CH_2-CH=CH-CH_2\sim$

$\sim CH_2-CH=CH-CH_2-CH_2-CH=CH-CH_2-CH_2-CH=CH-CH_2\sim$ (–S–S– crosslinks)

$\sim CH_2-CH=CH-CH_2-CH-CH-CH_2-CH_2-CH_2-CH=CH-CH\sim$ (–S–S– and –S–S–S– crosslinks)

$\sim CH_2-CH=CH-CH_2-CH_2-CH=CH-CH_2-CH-CH-CH_2-CH_2\sim$ (–S–S– crosslink)

Vulcanization of rubber

1.2.5 On the Basis of Behavior of Polymers Towards Temperature

Polymers can be classified into two categories on the basis of their behavior towards temperature:

Thermoplastics: They are those polymers which can be softened on heating and hardened on cooling, e.g., linear polymers. Since no chemical change occurs in the structure of the polymer molecule as a result of heating, therefore these polymers can be recycled many times.

Thermosets: They are those polymers which cannot be softened once they get solidified either by heating or by curing agents. Thermoset polymers cannot be recycled as a chemical change occurs in thermoset polymers as a result of heating. e.g., crosslinked epoxy resins, polydicyclopentadiene, polycarbonates, polyisoprene etc.

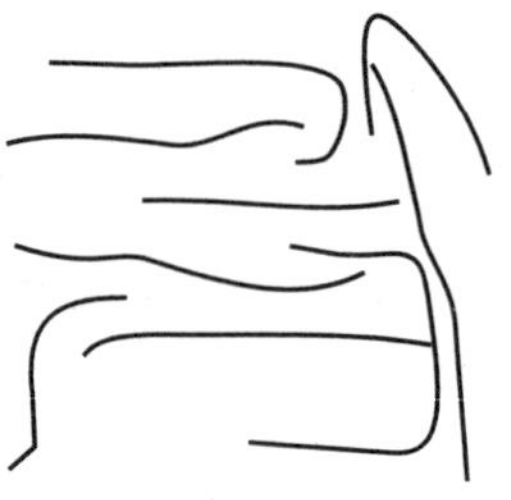

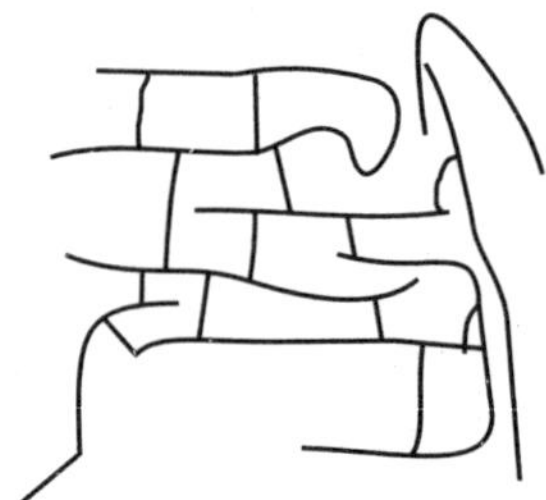

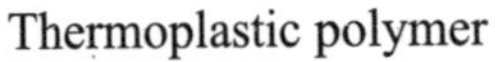

Thermoplastic polymer Thermoset polymer

1.2.6 On the Basis of Tacticity or the Stereochemistry of the Polymers

Optical Isomerism in Polymers

The orientation of the monomeric units in a polymer chain is called tacticity of the polymer. (The difference in configuration is called 'tacticity'.) When different substituents are linked to a carbon atom (chiral carbon) in a polymer molecule, it shows optical isomerism.

Polyethylene does not have any asymmetric carbon, therefore, it does not show any optical isomerism. In this structure if we put all the carbons on a plane and put hydrogen attached to it above the plane and other hydrogen attached to it below the plane. and then if a free rotation of carbon-carbon single bond occurs, there will be no change in the resultant spatial arrangement of the polymer segment because all the hydrogens attached to carbon atom are indistinguishable.

$$\begin{array}{ccccccccccccccccc} & & H & & H & & H & & H & & H & & H & & H & & H \\ & & | & & | & & | & & | & & | & & | & & | & & | \\ \sim\sim & - & C & - & C & - & C & - & C & - & C & - & C & - & C & - & C-\sim\sim \\ & & | & & | & & | & & | & & | & & | & & | & & | \\ & & H & & H & & H & & H & & H & & H & & H & & H \end{array}$$

But imagine that one of hydrogen atoms in all ethylene units of this polymer is substituted by a substituent like $-CH_3$, –Cl or –CN (–R) then the structural formula for the polymer will be as follows.

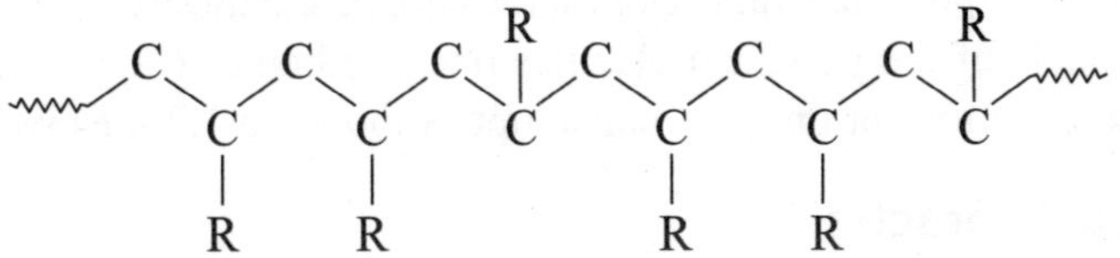

In this structure every alternate carbon atom (C*) is asymmetric (chiral) because C* has four different substituents attached to it. One H, one R,. one $-CH_2-$ of length *m* and other $-CH_2-$ of length *n*. Now each C* atom provides site for optical isomerism. Each site can exhibit either *R*- or *S*- type isomerism.

On the basis of optical activity, polymers can be divided into following types:

Atactic or Heterotactic : A polymer, in which there is no systematic configuration, is called an atactic polymer, e.g., polypropylene. The combination of two monomer units may take place either way – head to tail, tail to tail or head to head. This way we will have R groups located randomly in the polymer chain. Atactic polymers are generally easy to synthesise, but these polymers cannot crystallise due to their irregularity.

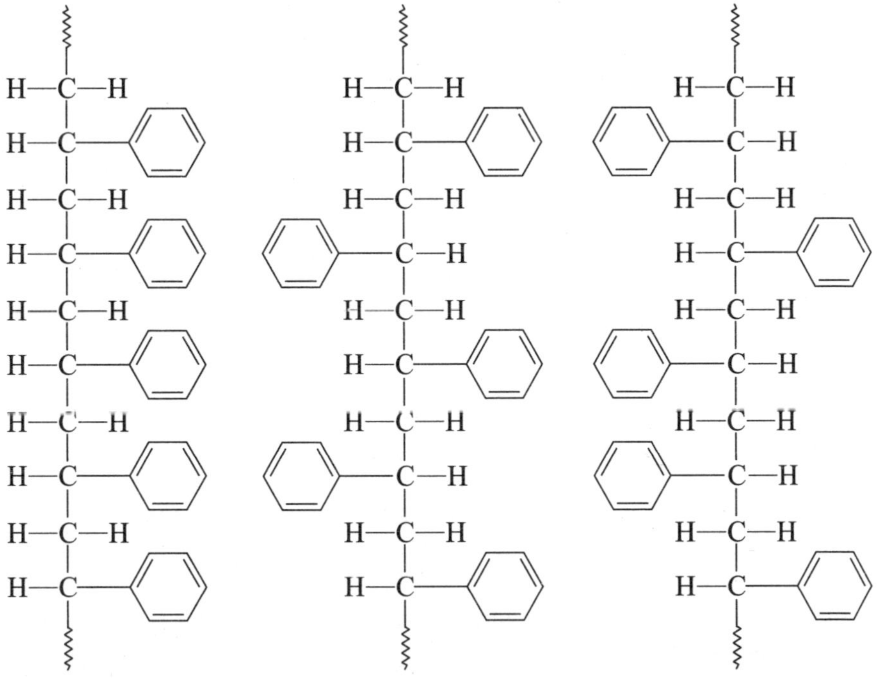

Isotactic Syndiotactic Atactic

Polystyrene

Isotactic: A polymer, in which substituent group is located on one side of plane of carbon-carbon chain, i.e., every chiral carbon has the the same configuration, is called isotactic.

Syndiotactic: A polymer, in which substituent group is attached alternatively above and below the plane of carbon-carbon chain, is called syndiotactic.

With suitable polymerization routes, it is possible to prepare isotactic and syndiotactic polymers (at least in some cases), and these can crystallise. Building up models of long chain molecules one can begin to see that there are many possible conformations they can adopt as shown in 1,2 -disubstituted ethylenes.

Threodiisotactic

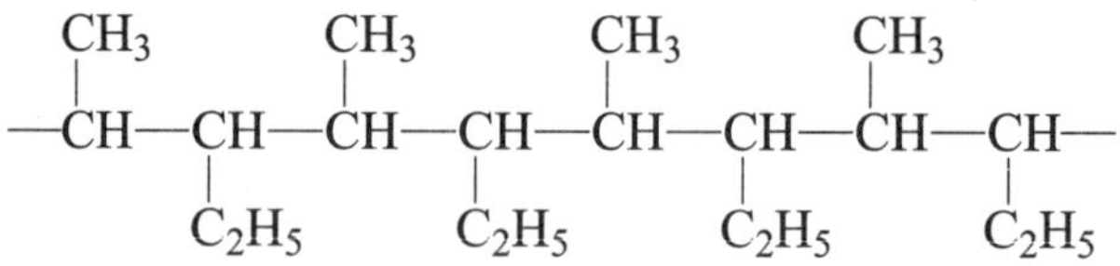

```
 CH3        CH3        CH3        CH3
  |          |          |          |
—CH—CH—CH—CH—CH—CH—CH—CH—
       |          |          |          |
      C2H5       C2H5       C2H5       C2H5
```

Erythrodiisotactic

```
 CH3 C2H5 CH3 C2H5 CH3 C2H5 CH3 C2H5
  |    |    |    |    |    |    |    |
—CH—CH—CH—CH—CH—CH—CH—CH—
```

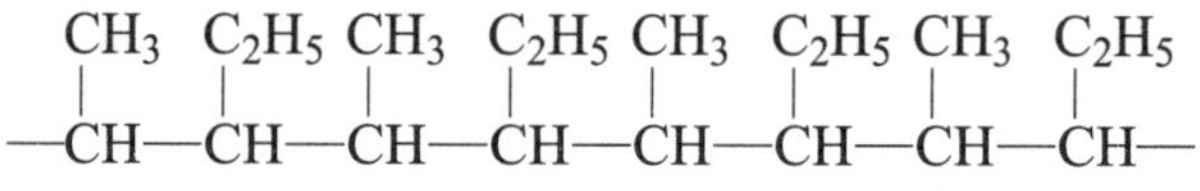

Erythrodisyndiotactic

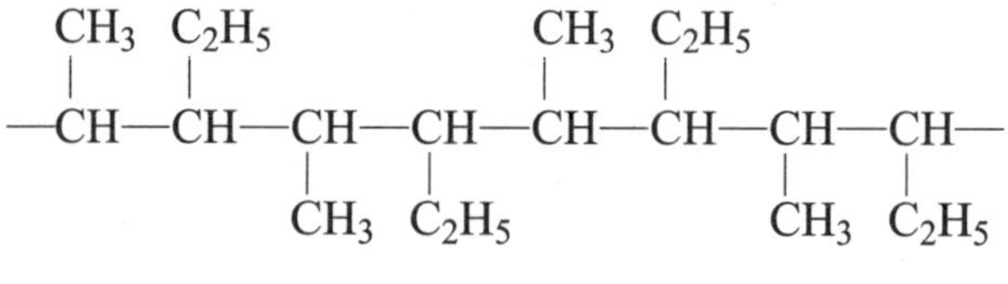

```
 CH3 C2H5              CH3 C2H5
  |    |                |    |
—CH—CH—CH—CH—CH—CH—CH—CH—
            |    |                |    |
           CH3 C2H5              CH3 C2H5
```

Threodisyndiotactic

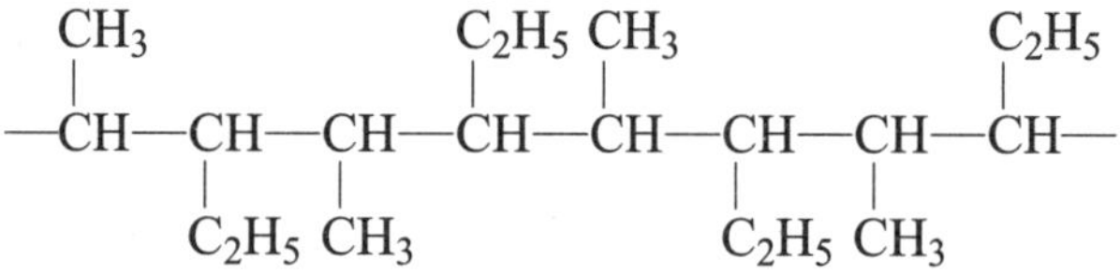

```
 CH3             C2H5 CH3             C2H5
  |                |    |                |
—CH—CH—CH—CH—CH—CH—CH—CH—
       |    |                |    |
      C2H5 CH3              C2H5 CH3
```

Geometrical Isomerism in Polymers

In optical isomerism C-C single bonds are involved but in geometrical isomerism C = C double bonds are involved. Geometric isomerism depends on how the substituents groups are positioned on a C = C double bond, e.g., 1,3-butadiene (synthetic rubber). Polymers having double bonds in their back bones show *cis* and *trans* isomerism.

Cis

Trans

1.2.7 On the Basis of Morphology

Polymers can be classified into two classes on the basis of morphology:

Crystalline Polymers: Invariably do not form perfect crystalline materials but instead are semicrystalline with both crystalline and amorphous regions. The crystalline phase of such polymers are characterized by their melting temperature (T_m).

Amorphous Polymers: Characterized by their glass transition temperature (T_g), the temperature at which they transform abruptly from the glass state (hard) to the rubbery state (soft). This transition corresponds to the onset of chain motion.

1.2.8 On the Basis of End Use

On the basis of their end use, polymers can be divided into three classes:

Rubbers: They are dimensionally unstable. These polymers are characterized by long-range elasticity. The rubbery polymers are characterized by low molecular cohesion. Their mechanical properties are poor. Their tensile strength ranges between 300–1000 psi.

Plastics: These are much stronger than rubbers. They exhibit tensile strength ranges between 4000–13000 psi. Some plastics are hard, stiff and dimensionally stable and some are soft and flexible.

Fibres: They are the strongest of the three different types of polymers. They exhibit tensile strength which ranges between 20000–150000 psi.

1.3 BONDING IN POLYMERS

Various types of bonding may exist in polymeric materials. These types are:

(i) Primary covalent
(ii) Hydrogen bonding
(iii) Dipole interaction
(iv) Van der Waals forces
(v) Ionic

Hydrogen bonding, dipole interaction, van der Waals, and ionic bondings are collectively known as secondary forces. Table 1.1 lists the approximate bond strengths of the bonds encountered in polymeric materials. These bonds break or dissociate when the available thermal energy exceeds their dissociation energy. The van der Waals bonds are the weakest and the primary covalent bonds are the strongest.

Table 1.1

Bond Type	***Dissociation Energy* (kcal/mol)**
Primary Covalent	50–200
Hydrogen bond	3–7
Dipole interaction	1.5–3
van der Waals	0.5–2
Ionic	10–20

(i) Primary Covalent Bond

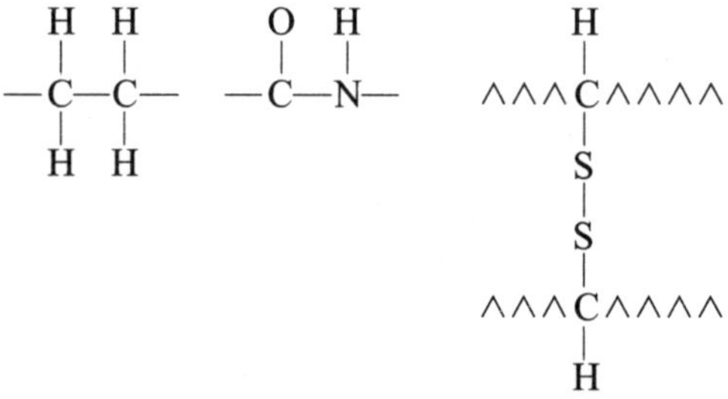

```
  H  H       O  H          H
  |  |       |  |          |
—C—C—     —C—N—     ^^^C^^^^
  |  |                     |
  H  H                     S
                           |
                           S
                           |
                       ^^^C^^^^
                           |
                           H
```

(ii) Secondary Bonding

(a) Hydrogen bond

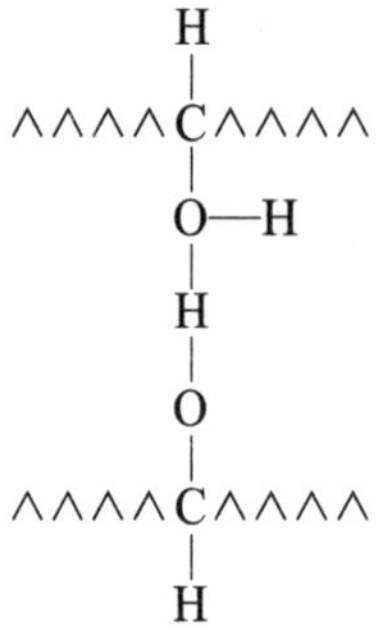

```
       H
       |
^^^^C^^^^
       |
       O—H
       |
       H
       |
       O
       |
^^^^C^^^^
       |
       H
```

(b) Dipole Interaction

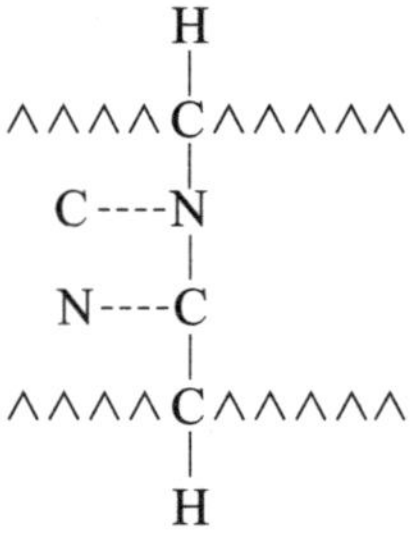

```
         H
         |
^^^^C^^^^^
         |
   C----N
         |
   N----C
         |
^^^^C^^^^^
         |
         H
```

(c) Ionic

```
         H
         |
^^^^C^^^^
         |
         C=O
         |
         O
         |
         Zn
         |
         O
         |
         C=O
         |
 ^^^C^^^^
         |
         H
```

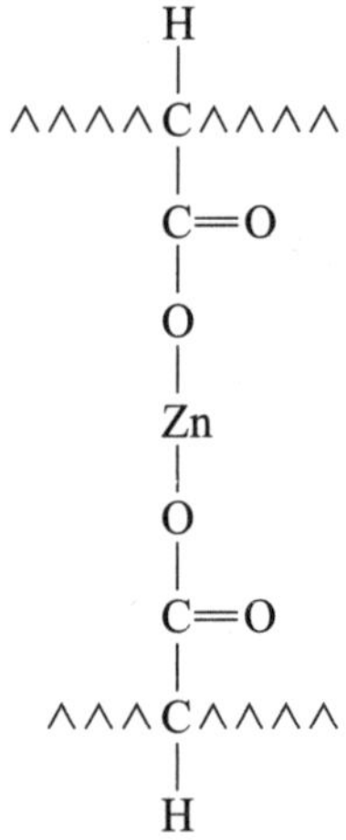

(d) Van der Waals

```
CH2     CH2     CH2     CH2
   \   /   \   /   \   /   \   /
    CH2     CH2     CH2     CH2

CH2     CH2     CH2     CH2
   \   /   \   /   \   /   \   /
    CH2     CH2     CH2     CH2
```

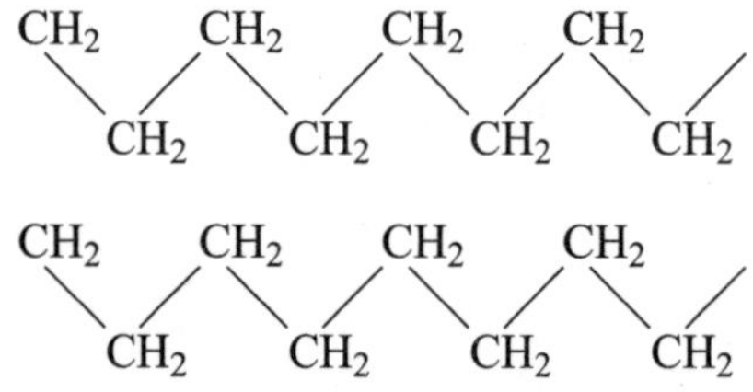

Relationship Between Bonding and Temperature

In linear and branched polymers only the secondary bonds hold the individual polymer chains together. Thus, as the temperature is raised, a point will be reached where the forces holding the chain together become insignificant, and they are then free to slide past one another, i.e., to flow upon the application of stress. Therefore linear and branched polymers are generally thermoplastic. The main chains in a crosslinked polymer, on the other hand, are held together by the same primary covalent bonds as the atoms in main chains. When the thermal energy exceeds the dissociation energy of the primary covalent bond, both main chain and crosslink bonds fall randomly, and the polymer degrades. Hence crosslinked polymers are thermosetting.

Relationship Between Bonding and the action of Solvent

The action of solvents on polymers is in many ways similar to that of heat. Appropriate solvents can penetrate, replace the interchain secondary bonds, and thereby pull apart and dissolve linear and branched polymers. The polymer-solvent secondary bonds cannot overcome primary valence crosslinks, however; crosslinked polymers are not soluble, although they may swell extensively. The amount of swelling is, in fact, a convenient measure of the extent of crosslinking. A lightly crosslinked polymer will swell tremendously, while one with extensive crosslinking, will not swell noticeably at all.

1.4 COPOLYMERS

In copolymers monomer units may be arranged regularly or at random. So, on the basis of arrangement of monomeric units, copolymers can be classified in the following types:

Statistical copolymers The copolymers in which the sequential distribution of the repeat units obeys known statistical laws.

Random copolymer A copolymer, in which the two monomer units are placed in random manners (irregularly), is termed as random copolymer.

—A—A—B—A—B—B—A—B—A—A—B—B—B—A—

Alternate copolymer A copolymer, in which the two monomer units are placed in an alternative manner, is called alternate copolymer.

—A—B—A—B—A—B—A—B—A—B—A—B—A—B—

Block copolymer It is a linear copolymer in which each monomer unit forms a block.

—A—A—A—A—A—A—A—B—B—B—B—B—B—B—

Graft copolymer It is a branched copolymer in which one monomer unit forms the backbone of the polymer and the other monomer unit forms the branches.

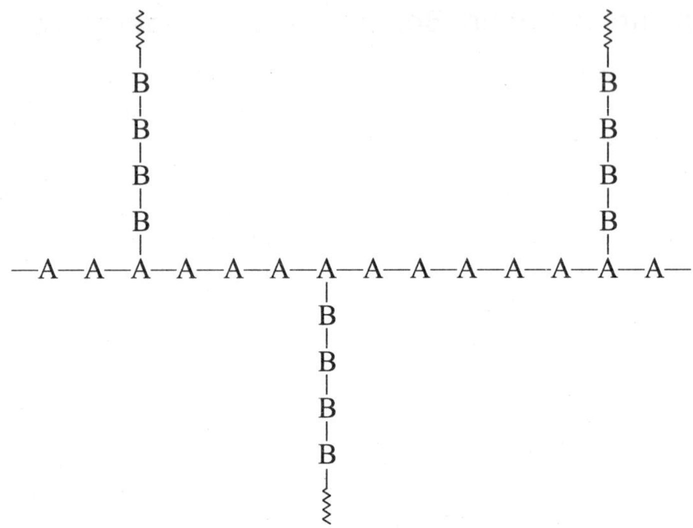

1.5 WHY POLYMERS ARE DIFFERENT FROM SMALL MOLECULES

Why polymers, the macromolecules act differently from small molecules. There are three ways in which polymers will act differently from small molecules. And the reasons are a little bit more complicated than just "because they are bigger". The three differences are usually named as:

(i) Chain entanglement
(ii) Summation of intermolecular forces
(iii) Time scale of motion

Chain Entanglement

Remember now that most polymers are linear polymers; i.e., they are molecules whose atoms are joined in a long line to form a huge chain. Now most of the time, but not always, this chain is not stiff and straight, but is flexible. It twists and bends around to form a tangled mess. The chains tend to twist and wrap around each other, so the polymer molecules collectively will form one huge tangled mess.

Now when a polymer is molten, the chains will act like spaghetti angled up on a plate. If one tries to pull out any one strand of spaghetti, it slides right out with no problem. But when polymers are cold and in the solid state, they act more like a tangled up old ball of string. Trying to pull one strand out of this mess is a little harder and one is more likely to end up making a big knot. Solid polymers are like this. The chains are all tangled up in each other and it is difficult to untangle them. This is what makes so many polymers strong in materials like plastics, paint, elastomers, and composites.

Summation of Intermolecular Forces

All molecules, both small ones and polymers, interact with each other, attracting each other through electrostatics. Some molecules are drawn to each other more than others. Polar molecules stick together better than nonpolar molecules. For example, water and methane have similar molecular weights. Methane's weight is sixteen and water's is eighteen. Methane is a gas at room temperature, and water is a liquid. This is because water is very polar, polar enough to stick together as a liquid, while metane is very nonpolar, so it does not stick together very well at all. Intermolecular forces affect polymers just like small molecules but with polymers, these forces are greatly compounded. The bigger the molecule, the more molecule there is to exert an intermolecular force. Even when only weak van der Waals forces are at play, they can be very strong in binding different polymer chains together. This is another reason why polymers can be very strong as materials. Polyethylene, for example, is very nonpolar. It only has van der Waals forces to play with, but it is so strong that it is used to make bullet proof vests.

Time Scale of Motion

A bunch of small molecules can move around a lot faster and a lot more chaotically when they are not all tied to each other. Tie the molecules together in a big long chain and they slow down. So then how does this make a polymeric material different from a material made of small molecules? This slow speed of motion makes polymers do some very unusual things. For one, if a polymer is dissolved in a solvent, the solution will be a lot more viscous than the pure solvent.

1.6 FLOWCHART OF POLYMERIC MATERIALS

Inorganic

- **Natural**
  - Clay
    - Bricks
    - Cement
    - Pottery
  - Sand
    - Glass
- **Synthetic**
  - Fibers

Organic

- **Synthetic**
  - Adhesives
  - Fibers
  - Coatings
  - Plastics
  - Rubbers
- **Natural**
  - Polysaccharides
    - Adhesives
    - Fibers
  - Proteins
    - Adhesives
    - Fibers
  - Polyisoprene
    - Rubbers

2

CHAIN GROWTH POLYMERIZATION

Learning Objectives

After studying this chapter, you will know:

- What are addition polymers?
- What various modes are used for synthesizing addition polymers?
- What is the reaction mechanism for radical and non-radical type addition polymerization?
- What are the methods to prepare stereoregular polymers?

Chapter Contents

• Polymerization • Free radical addition • Non-radical addition polymerization • Cationic polymerization • Anionic polymerization • Heterogeneous stereospecific polymerization • Coordination polymerization • Metallocene and metathesis polymerization

A chemical reaction that makes polymers is called a polymerization. There are many of these reactions, and they come in all kinds. But all polymerizations have one thing in common: they all start with small molecules, which join into big giant molecules. We call the original small molecules monomers. But they can join in many different ways. These different ways are broadly classified as chain growth and step growth polymerization reactions. This chapter is devoted to the synthesis of polymers through chain growth polymerization and the reaction kinetics.

In a chain growth polymerization, monomers become part of the polymer one at a time. Chain growth polymerizations are usually addition polymerization but they can also be condensation polymerizations.

2.1 ADDITION POLYMERIZATION

In this type of polymerization, the monomers add on to one another in such a way that the polymer formed contains all the atoms of the starting monomers. In other words, the molecular weight of an addition polymer is a multiple of the molecular weight of the monomer units. For example, the formation of polyethylene from ethylene.

$$\underset{\text{Ethylene}}{nCH_2{=}CH_2} \longrightarrow \underset{\text{Polyethylene}}{-CH_2{-}CH_2{-}[-CH_2{-}CH_2{-}]{-}CH_2{-}CH_2{-}}$$

The addition polymerization is encountered in unsaturated compounds such as all the vinyl and acrylic compounds.

There are three types of reaction intermediates for addition polymerization:

(i) ($C^{\cdot}$) free radical
(ii) (C^{+}) carbocation
(iii) (C^{-}) carbanion

2.1.1 Free Radical Addition

One of the most important types of addition polymerization is initiated by the action of free radicals, electrically neutral species with an unshared electron. Free radicals for the initiation of addition polymerization are usually generated by the photo or the thermal decomposition of organic peroxides or azo-compounds.

$$C_3H{-}C(CH_3)(CN){-}N{=}N{-}C(CH_3)(CN){-}CH_3$$

2-2′-azo-bis-isobutyrylnitrile (AIBN)

$$C_6H_5{-}C(=O){-}O{-}O{-}C(=O){-}C_6H_5$$

Benzoyl peroxide (BPO)

Free radical formation by dissociation of AIBN and BPO

$$H_3C{-}C(CH_3)(CN){-}N{=}N{-}C(CH_3)(CN){-}CH_3 \longrightarrow H_3C{-}\dot{C}(CH_3)(CN) + N{\equiv}N + \cdot C(CH_3)(CN){-}H_3C$$

$$C_6H_5{-}C(=O){-}O{-}O{-}C(=O){-}C_6H_5 \longrightarrow C_6H_5{-}C(=O){-}O\cdot + \cdot O{-}C(=O){-}C_6H_5$$

$$C_6H_5{-}C(=O){-}O\cdot \longrightarrow C_6H_5\cdot + O{=}C{=}O$$

Reaction Mechanism

Initiator(I) decomposition:

$$\dot{\mathrm{I}} \xrightarrow{K_d} 2\dot{\mathrm{R}} \tag{2.1}$$

Followed by chain initiation:

$$\dot{\mathrm{R}} + \mathrm{M} \xrightarrow{K_i} \mathrm{R}\dot{\mathrm{M}}_1 \tag{2.2}$$

Chain propagation

$$\left.\begin{array}{l} \mathrm{R}\dot{\mathrm{M}}_1 + \mathrm{M} \xrightarrow{K_p} \mathrm{R}\dot{\mathrm{M}}_2 \\ \mathrm{R}\dot{\mathrm{M}}_{n-1} + \mathrm{M} \xrightarrow{K_p} \mathrm{R}\dot{\mathrm{M}}_n \end{array}\right\} \tag{2.3}$$

Where, I is the initiator molecule, R is the radical formed by the decomposition of initiator, RM_1 is the growing radical and M is the monomer molecule.

Chain termination can occur by

(i) coupling of two growing polymer chain
(ii) disproportionation of chains
(iii) chain transfer by such molecules that can generate two radicals, one of which is quite stable and the other is quite unstable, e.g., thioalcohols (R—SH)

$$\mathrm{R}\dot{\mathrm{M}}_m + \mathrm{R}\dot{\mathrm{M}}_n \xrightarrow{K_{t_c}} \mathrm{R} - \mathrm{M}_{m+n} - \mathrm{R} \quad \text{by coupling} \tag{2.4}$$

$$\mathrm{R}\dot{\mathrm{M}}_m + \mathrm{R}\dot{\mathrm{M}}_n \xrightarrow{k_{t_d}} \mathrm{RM}_n + \mathrm{M}_n \quad \text{by disproportionation} \tag{2.5}$$

$$\mathrm{RM}_m + \mathrm{SH} \longrightarrow \mathrm{RM}_m + \mathrm{S} \quad \text{by chain transfer} \tag{2.6}$$

Polyethylene Formation by Free Radical Mechanism

Initiation:

$$C_6H_5\cdot + CH_2{=}CH_2 \longrightarrow C_6H_5{-}CH_2{-}CH_2\cdot$$

Propagation:

$$C_6H_5{-}CH_2{-}CH_2\cdot + CH_2{=}CH_2 \longrightarrow$$

Termination:

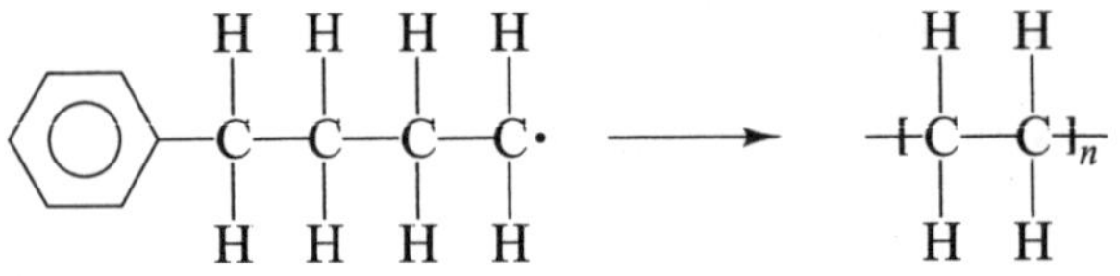

Termination by coupling:

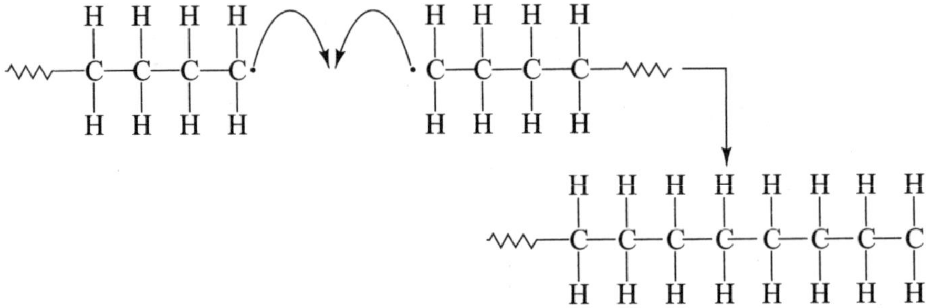

Termination by disproportionation:

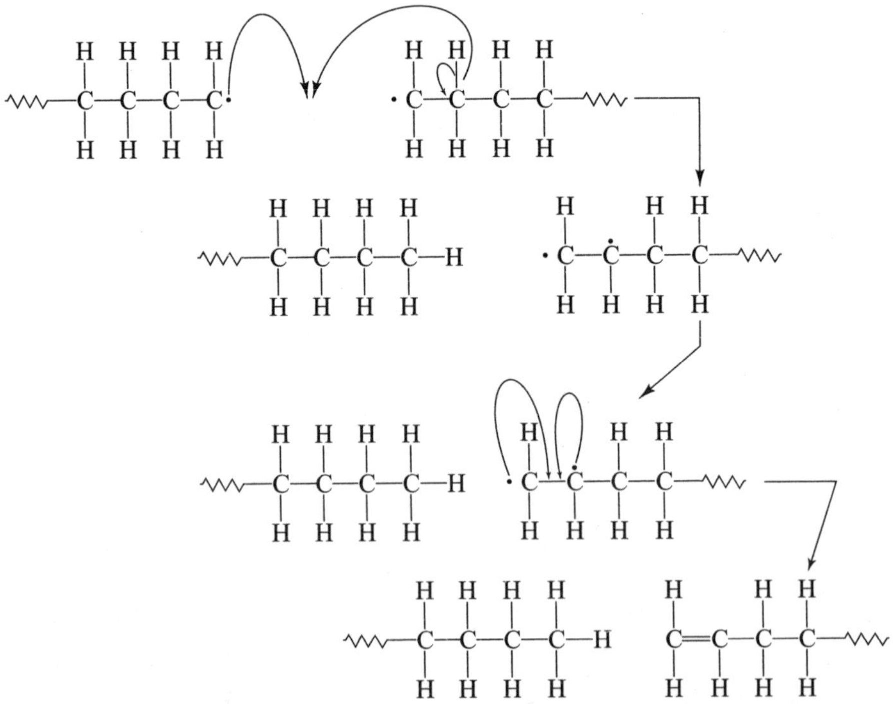

In disproportionation, when two growing-chain ends come close together, the unpaired electron of one chain does something strange. Rather than simply joining with the unpaired electron of the other chain, it finds one in the carbon-

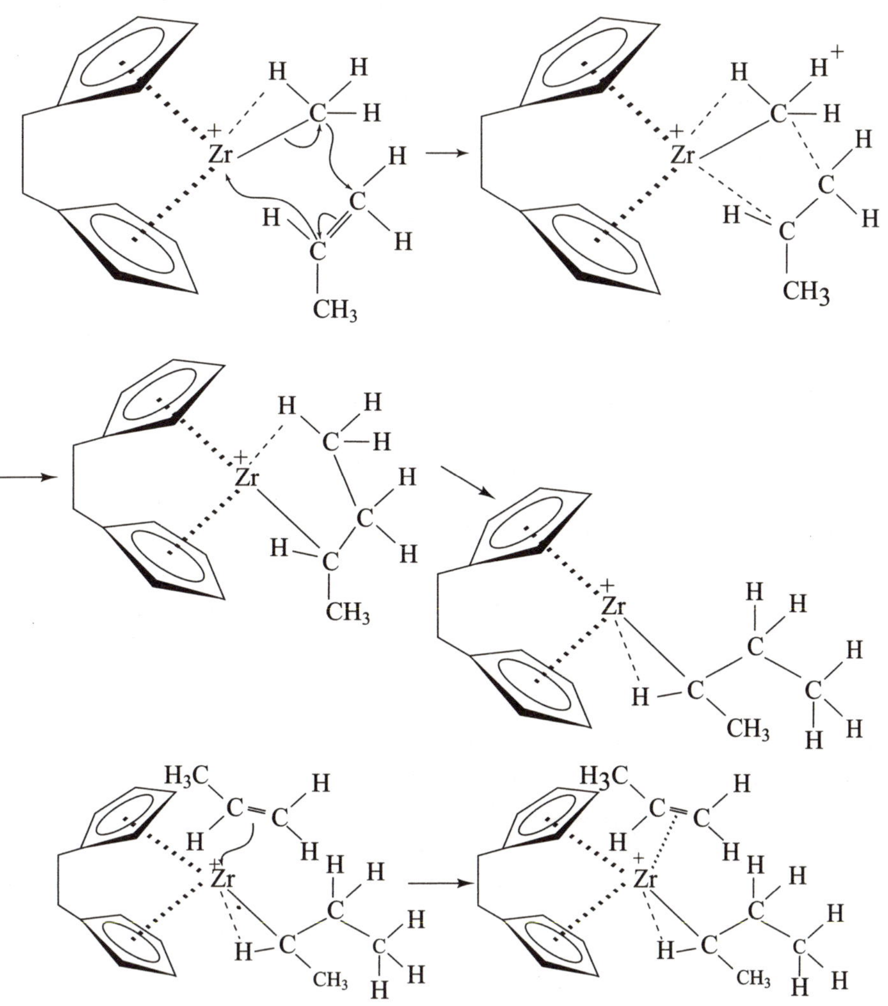

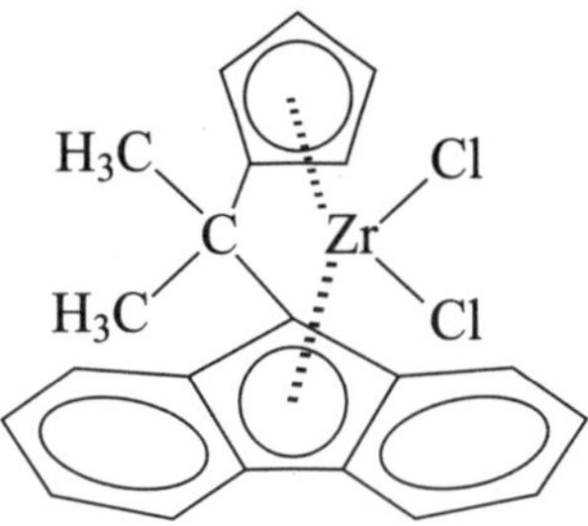

Ring Opening Metathesis Polymerization

Polymers may be formed by cyclic olefins, too. A cyclic olefin, cyclopentene, is used to make a polymer which does not have cyclic structures in its backbone. This is why it is called as "ring-opening metathesis polymerization". A nifty little molecule, norbornene, is polymerized by ROMP to get polynorbornene. It is a kind of rubber used to make auto parts.

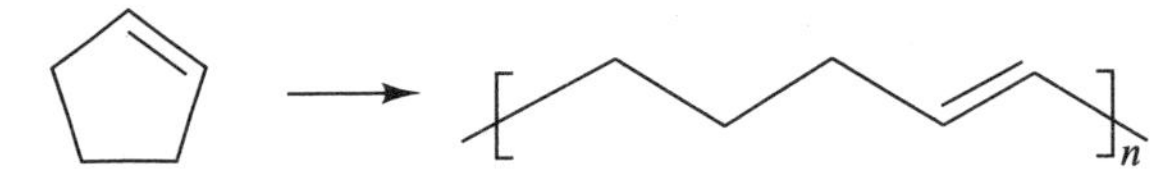

Cyclopentene

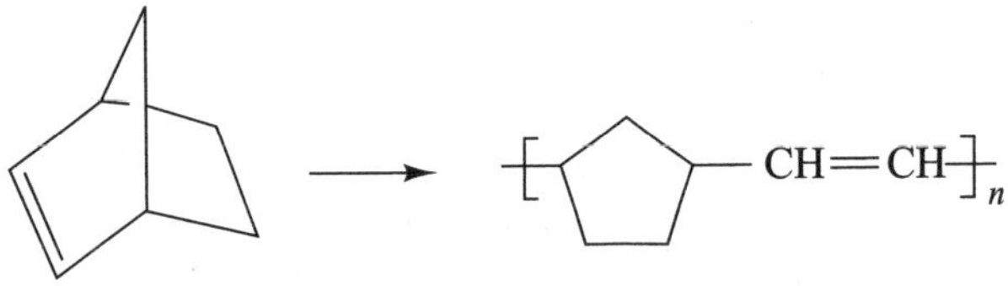

$-\!\!\left[\ \bigcirc\!-\!CH{=}CH\ \right]_n$

Norbornene Polynorbornene

A polymer with a cyclic olefin in a pendant group can also be made using ROMP. For example, polydicyclopentadiene, is prepared from endo-dicyclopentadiene.

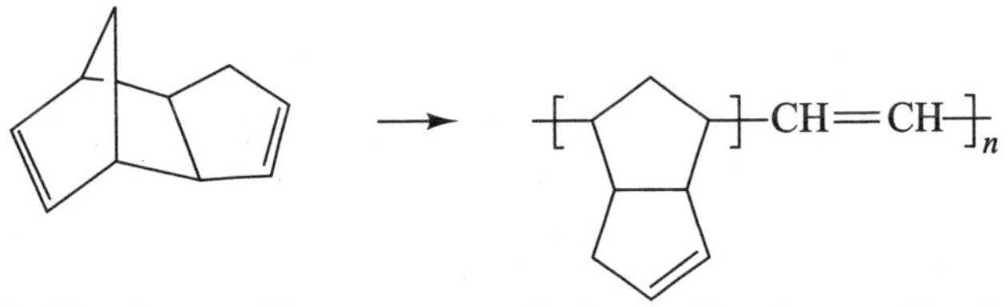

$-\!\!\left[\ \ \right]\!-CH{=}CH-\!\!\right]_n$

endo-dicyclopentadiene poly (*endo-dicyclopentadiene*)

$-\!\!\left[\ \ \right]\!-CH{=}CH-\!\!\right]_n$

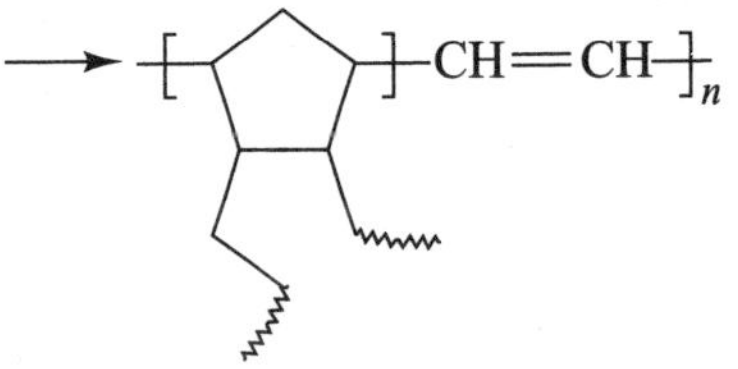

Crosslinked polymer

This pendant olefin can be reacted to crosslink the polymer. This polymer is a hard plastic that is used to make satellite dish antennae, body parts for snowmobiles, and other things that are generally used in the outside exposed to environment.

Reaction Mechanism

The initiator is usually a compound with a carbon atom double bonded to a metal atom, like tungsten or molybdenum. The initiator reacts with a molecule of cyclopentene in a metathesis reaction to get a molecule with a carbon-carbon double bond at one end, and a carbon-metal double bond at the other.

The carbon-metal double bond can react with another molecule of cyclopentene in the exact same way.

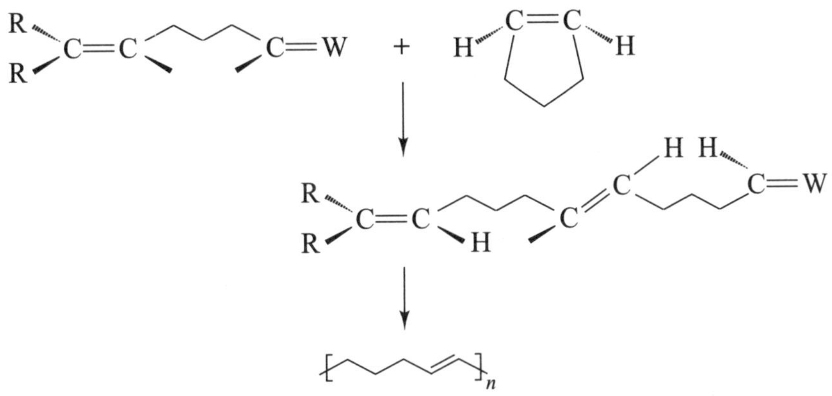

2.3 CONDENSATION POLYMERIZATION

There are also condensation polymerizations that are chain growth polymerizations.

$$\mathrm{Cl{-}C({=}O){-}C_6H_4{-}C({=}O){-}O{-}CH_2{-}CH_2{-}OH + HO{-}CH_2{-}CH_2{-}OH} \longrightarrow$$

$$\mathrm{HO{-}CH_2{-}CH_2{-}O{-}C({=}O){-}C_6H_4{-}C({=}O){-}O{-}CH_2{-}CH_2{-}OH + HCl}$$

$$\mathrm{Cl{-}C({=}O){-}C_6H_4{-}C({=}O){-}O{-}CH_2{-}CH_2{-}CH_2{-}OH + Cl{-}C({=}O){-}C_6H_4{-}C({=}O){-}O{-}CH_2{-}CH_2{-}OH} \longrightarrow$$

$$\mathrm{Cl{-}C({=}O){-}C_6H_4{-}C({=}O){-}O{-}CH_2{-}CH_2{-}O{-}C({=}O){-}C_6H_4{-}C({=}O){-}O{-}CH_2{-}CH_2{-}OH + HCl}$$

It can do other things, too. It may react with another dimer to form a tetramer or it may even react with a trimer to form a pentamer. Tetramers and pentamers can react to form even bigger oligomers. And so they continue to grow until eventually the oligomers are big enough to be called polymers. In a step growth reaction, the growing chains may react with each other to form even longer chains. This applies to chains of all lengths. The monomer or dimer may react in just the same way as a chain hundreds of monomer units long. But in an addition ploymerization, only monomers react with growing chains.

3.2 NON-CONDENSATION POLYMERIZATION

There are addition polymerizations that are step growth polymerizations. Polymerization may occur by the ionic addition of monomers or by the oxidative coupling. The mechanisms of these polymerizations may be any of addition polymerization but the kinetics are distinctly step-growth, which means that high molecular weight polymer is obtained only at the end of polymerization. One example is the polymerization that produces polyurethanes.

$$\mathrm{O{=}C{=}N{-}C_6H_4{-}CH_2{-}C_6H_4{-}N{=}C{=}O + HO{-}CH_2{-}CH_2{-}OH}$$

4, 4-diisocyanatodiphenylmethane Ethylene glycol

$$\longrightarrow \mathrm{{-}[C({=}O){-}N(H){-}C_6H_4{-}CH_2{-}C_6H_4{-}N(H){-}C({=}O){-}O\ CH_2{-}CH_2{-}O]_n{-}}$$

Polyurethane

a four-membered transition state and the zirconium ends up just like it started, lacking a ligand, but with an agostic association with a C-H bond from the propylene monomer. Another propylene monomer can come along and react just like the first one did as shown below.

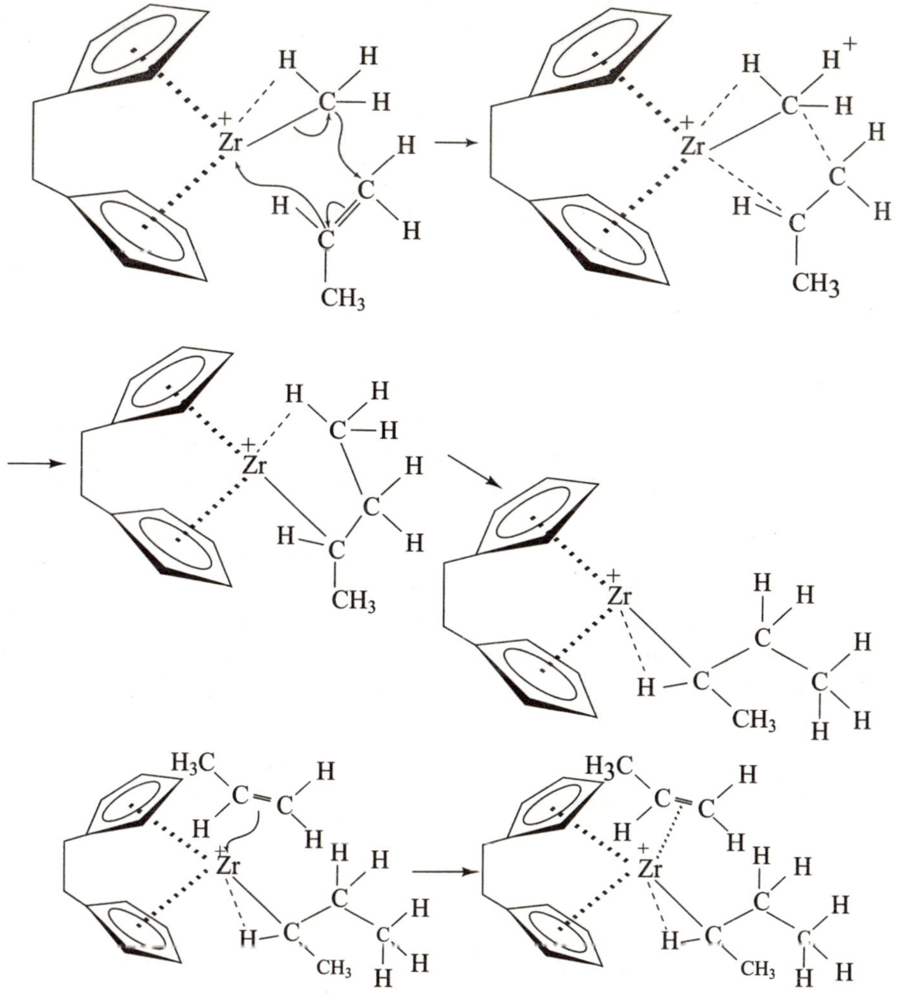

The propylene coordinates with the zirconium and then the electrons shuffle.

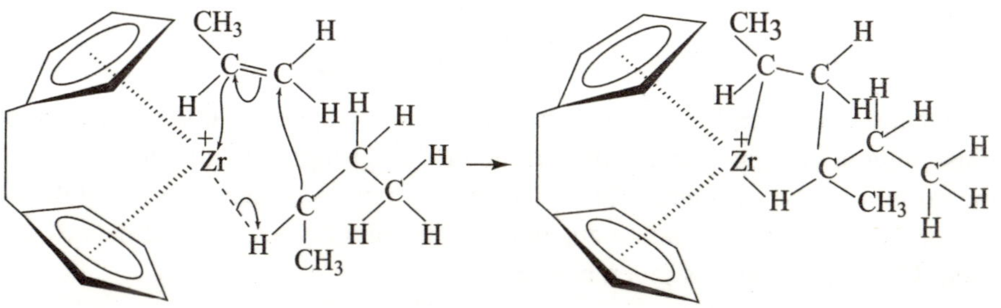

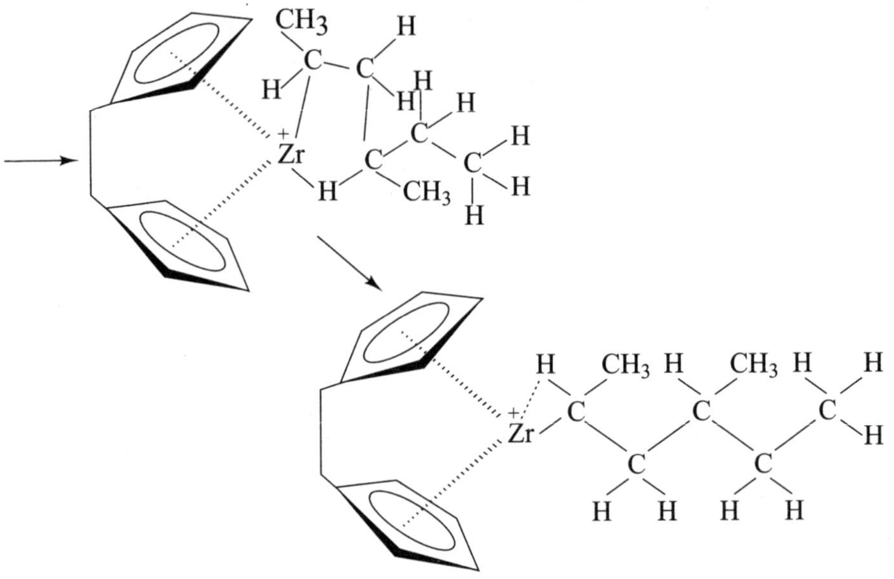

An isotactic polypropylene is formed this way as the propylene monomer always approaches the catalyst with its methyl group pointed away from the indenly ligand.

If the methyl group were pointed toward the indenyl ligand, the two would bump into each other, keeping the propylene from getting close enough to the zirconium to form a complex. So, only when the methyl group is pointed away from the indenyl ligand can the propylene complex be formed with zirconium. When the second monomer is added, it has to approach from the other side, and it also has to point its methyl group away from indenyl ring. Notice that this means the methyl group is pointed up rather than down.

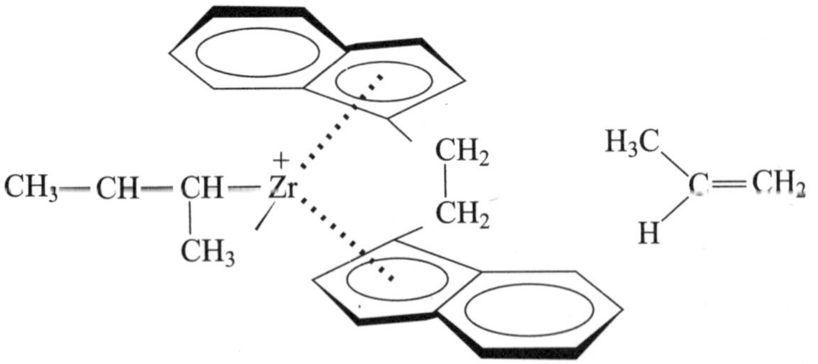

Because the second propylene is adding from the opposite side as the first, it must point in the opposite direction if the methyl groups are going to end up on the same side of the polymer chain. This catalyst gives isotactic polypropylene.

For the synthesis of syndiotactic polyolefins, other types of metallocenes are used. One of the them is given below.

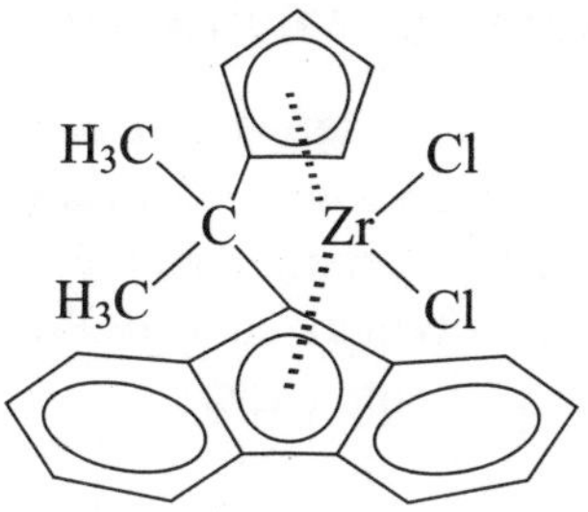

isopropyl (η^5-cyclopentadienyl-η^3-fluorenyl) ziroconium dichloride

Successive monomers approach from opposite sides of this catalyst, but they are always pointing their methyl groups up. This way, the methyl groups end up on alternating sides of the polymer chain.

2.2.3.3 Olefin Metathesis

Polymers with double bonds in the backbone chain are called polyalkenamers. These are formed by monomers with double bonds. One way to make polyalkenamers to use a nifty reaction called olefin metathesis. The two different olefin metathesis polymerizations are ring opening metathesis polymerization, or ROMP, and acyclic diene metathesis polymerization, or ADMET.

Acyclic Diene Metathesis Polymerization

In ADMET, polymerization is started with an acyclic diene, such as 1,5-hexadiene, and end up with a polymer with a double bond in the backbone chain, and by-product ethylene gas.

1,5-hexadiene $+ CH_2{=}CH_2$

Metathesis takes place by two molecules of 1,5-hexadiene reacting together as shown below.

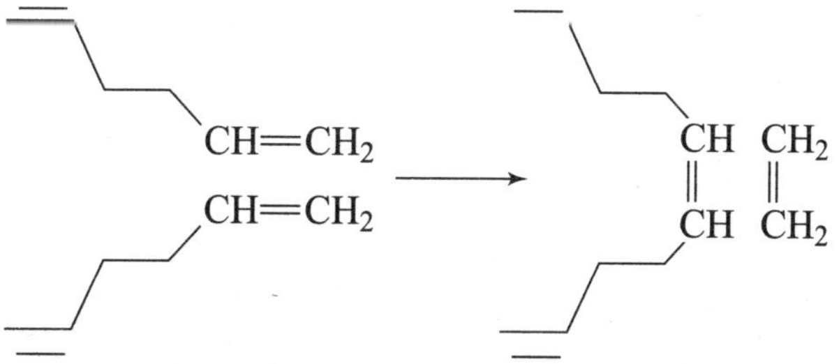

$$CH{=}CH_2 \quad CH{=}CH_2 \longrightarrow CH{=}CH \quad CH_2{=}CH_2$$

And of course the double bonds at the ends of the new molecule can have a little metathesis magic done on them as well, and in this way the polymer grows. Both *cis-* and *trans-* isomers are formed usually by this type of polymerization.

Ring Opening Metathesis Polymerization

Polymers may be formed by cyclic olefins, too. A cyclic olefin, cyclopentene, is used to make a polymer which does not have cyclic structures in its backbone. This is why it is called as "ring-opening metathesis polymerization". A nifty little molecule, norbornene, is polymerized by ROMP to get polynorbornene. It is a kind of rubber used to make auto parts.

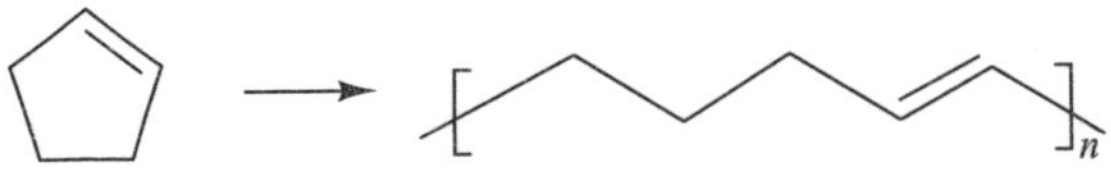

Cyclopentene

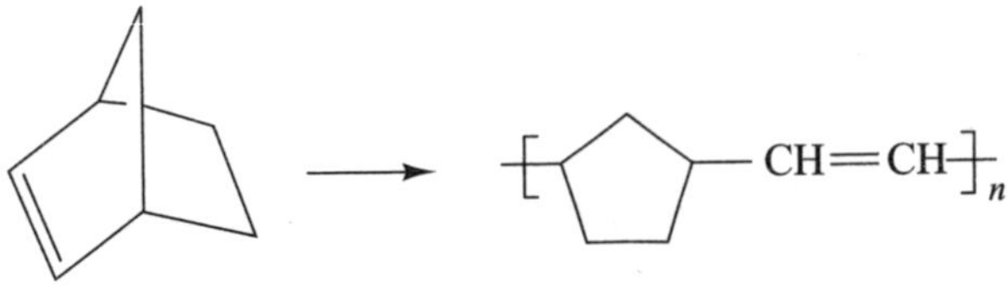

CH=CH n

Norbornene Polynorbornene

A polymer with a cyclic olefin in a pendant group can also be made using ROMP. For example, polydicyclopentadiene, is prepared from endo-dicyclopentadiene.

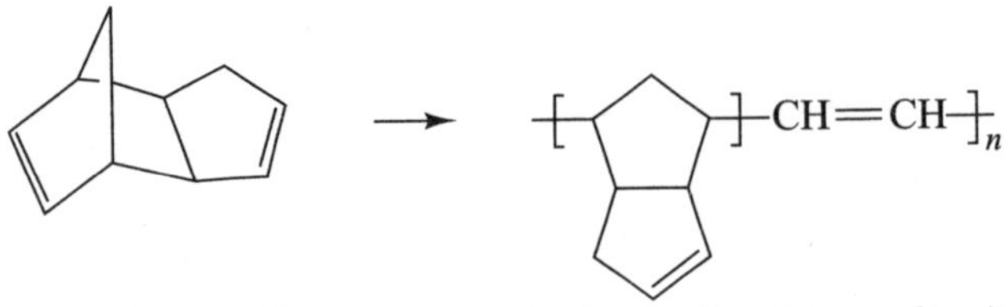

CH=CH n

endo-dicyclopentadiene poly (*endo-dicyclopentadiene*)

CH=CH n

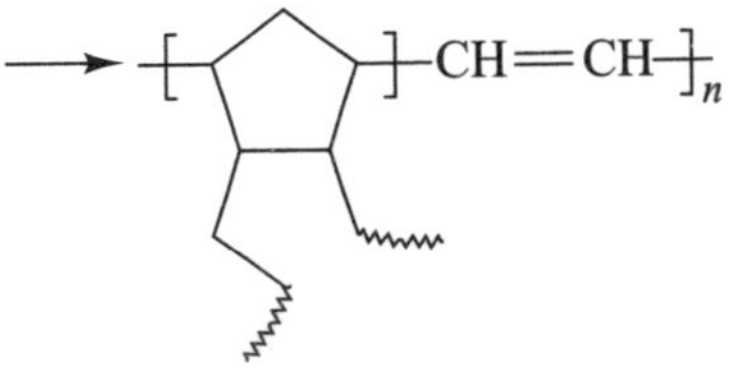

Crosslinked polymer

This pendant olefin can be reacted to crosslink the polymer. This polymer is a hard plastic that is used to make satellite dish antennae, body parts for snowmobiles, and other things that are generally used in the outside exposed to environment.

Reaction Mechanism

The initiator is usually a compound with a carbon atom double bonded to a metal atom, like tungsten or molybdenum. The initiator reacts with a molecule of cyclopentene in a metathesis reaction to get a molecule with a carbon-carbon double bond at one end, and a carbon-metal double bond at the other.

The carbon-metal double bond can react with another molecule of cyclopentene in the exact same way.

R
C=C
C=W + H
C=C
H
R
H H
R
C=C
C
C=W
R
H
C
n

2.3 CONDENSATION POLYMERIZATION

There are also condensation polymerizations that are chain growth polymerizations.

3

STEP GROWTH POLYMERIZATION

Learning Objectives

After studying this chapter, you will know:

- What are step-growth polymers?
- What is condensation polymerization?
- What is non-condensation polymerization?
- What are the prerequisites for the monomers to go under step-growth polymerization?

Chapter Contents

• Step-growth polymers • Condensation polymerization • Non-condensation polymerization • Functionality principle • Kinetics of reaction

In this classification scheme, a step growth polymerization is defined as the one that involves a random reaction of two molecules that may be any combination of a monomer, oligomer or a long chain molecule. High molecular weight polymer is formed only near the end of polymerization when most of the monomer has been depleted.

3.1 CONDENSATION POLYMERIZATION

Most step growth polymerizations involve a classical condensation reaction such as esterification, ester interchange, or amidation.

Amidation:

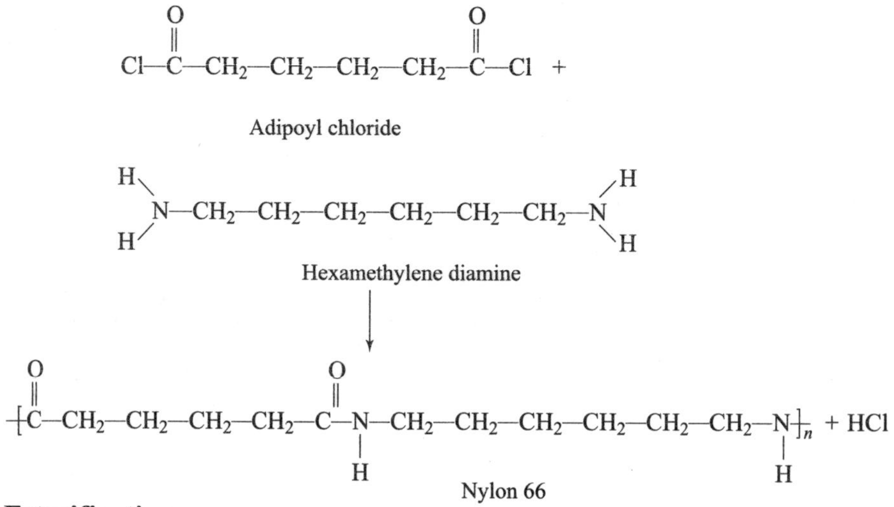

$$Cl-\overset{O}{\overset{\|}{C}}-CH_2-CH_2-CH_2-CH_2-\overset{O}{\overset{\|}{C}}-Cl \; +$$

Adipoyl chloride

$$\begin{matrix} H \\ H \end{matrix} \!\!> N-CH_2-CH_2-CH_2-CH_2-CH_2-CH_2-N <\!\! \begin{matrix} H \\ H \end{matrix}$$

Hexamethylene diamine

$$\downarrow$$

$$-\!\!\left[\overset{O}{\overset{\|}{C}}-CH_2-CH_2-CH_2-CH_2-\overset{O}{\overset{\|}{C}}-\underset{H}{\underset{|}{N}}-CH_2-CH_2-CH_2-CH_2-CH_2-CH_2-\underset{H}{\underset{|}{N}}\right]\!\!-_n + HCl$$

Nylon 66

Esterification:

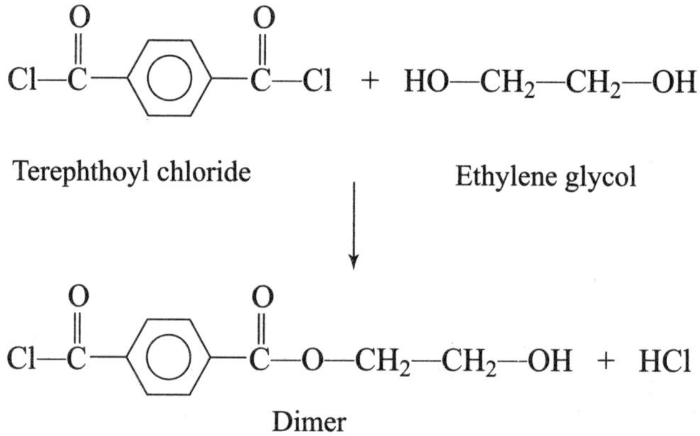

$$Cl-\overset{O}{\overset{\|}{C}}-C_6H_4-\overset{O}{\overset{\|}{C}}-Cl \; + \; HO-CH_2-CH_2-OH$$

Terephthoyl chloride Ethylene glycol

$$\downarrow$$

$$Cl-\overset{O}{\overset{\|}{C}}-C_6H_4-\overset{O}{\overset{\|}{C}}-O-CH_2-CH_2-OH \; + \; HCl$$

Dimer

Dimer can react with a molecule of terephthoyl chloride,

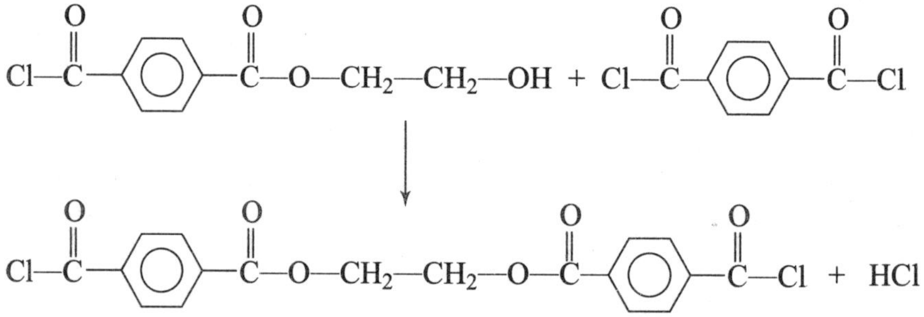

$$Cl-\overset{O}{\overset{\|}{C}}-C_6H_4-\overset{O}{\overset{\|}{C}}-O-CH_2-CH_2-OH + Cl-\overset{O}{\overset{\|}{C}}-C_6H_4-\overset{O}{\overset{\|}{C}}-Cl$$

$$\downarrow$$

$$Cl-\overset{O}{\overset{\|}{C}}-C_6H_4-\overset{O}{\overset{\|}{C}}-O-CH_2-CH_2-O-\overset{O}{\overset{\|}{C}}-C_6H_4-\overset{O}{\overset{\|}{C}}-Cl \; + \; HCl$$

or with a molecule of ethylene glycol.

$$Cl-\overset{O}{\overset{\|}{C}}-C_6H_4-\overset{O}{\overset{\|}{C}}-O-CH_2-CH_2-OH + HO-CH_2-CH_2-OH$$

$$\downarrow$$

$$HO-CH_2-CH_2-O-\overset{O}{\overset{\|}{C}}-C_6H_4-\overset{O}{\overset{\|}{C}}-O-CH_2-CH_2-OH + HCl$$

$$Cl-\overset{O}{\overset{\|}{C}}-C_6H_4-\overset{O}{\overset{\|}{C}}-O-CH_2-CH_2-CH_2-OH + Cl-\overset{O}{\overset{\|}{C}}-C_6H_4-\overset{O}{\overset{\|}{C}}-O-CH_2-CH_2-OH$$

$$\downarrow$$

$$Cl-\overset{O}{\overset{\|}{C}}-C_6H_4-\overset{O}{\overset{\|}{C}}-O-CH_2-CH_2-O-\overset{O}{\overset{\|}{C}}-C_6H_4-\overset{O}{\overset{\|}{C}}-O-CH_2-CH_2-OH + HCl$$

It can do other things, too. It may react with another dimer to form a tetramer or it may even react with a trimer to form a pentamer. Tetramers and pentamers can react to form even bigger oligomers. And so they continue to grow until eventually the oligomers are big enough to be called polymers. In a step growth reaction, the growing chains may react with each other to form even longer chains. This applies to chains of all lengths. The monomer or dimer may react in just the same way as a chain hundreds of monomer units long. But in an addition ploymerization, only monomers react with growing chains.

3.2 NON-CONDENSATION POLYMERIZATION

There are addition polymerizations that are step growth polymerizations. Polymerization may occur by the ionic addition of monomers or by the oxidative coupling. The mechanisms of these polymerizations may be any of addition polymerization but the kinetics are distinctly step growth, which means that high molecular weight polymer is obtained only at the end of polymerization. One example is the polymerization that produces polyurethanes.

$$O=C=N-C_6H_4-CH_2-C_6H_4-N=C=O + HO-CH_2-CH_2-OH$$

4, 4-diisocyanatodiphenylmethane — Ethylene glycol

$$\downarrow$$

$$\left[\overset{O}{\overset{\|}{C}}-\underset{H}{\underset{|}{N}}-C_6H_4-CH_2-C_6H_4-\underset{H}{\underset{|}{N}}-\overset{O}{\overset{\|}{C}}-O\ CH_2-CH_2-O\right]_n$$

Polyurethane

3.3 FUNCTIONALITY PRINCIPLE

According to the functionality principle and concept, for polymerization to take place, each of the reacting molecules must be at least bifunctional, giving average functionality for the system as two. For reacting compounds containing two different kinds of functional groups, the average functionality is calculated on the basis of stoichiometric equivalence of the two kinds of functional groups. For example, to strike stoichiometric equivalence of –COOH and –OH groups for an esterification reaction between a dicarboxylic acid and a trihydric alcohol, the acid and the alcohol should be taken in a molar ratio 3:2 which would express the average functionality for the system as 12/5 i.e., 2.4.

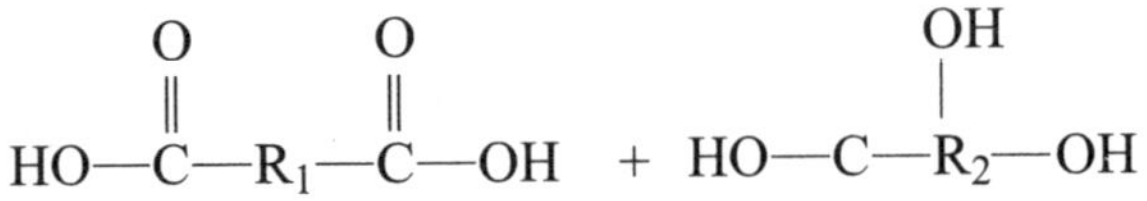

$$HO-\overset{\overset{\displaystyle O}{\|}}{C}-R_1-\overset{\overset{\displaystyle O}{\|}}{C}-OH \; + \; HO-\overset{\overset{\displaystyle OH}{|}}{C}-R_2-OH$$

3 moles of diacid + 2 moles of trihydric alcohol

$3 \times 2 = 6$ $\qquad$ $2 \times 3 = 6$

$6 + 6 = 12$ and $3 : 2 = 5$ so $12/5 = 2.4$ is the average functionality

For reaction between monohydric alcohol and triacid

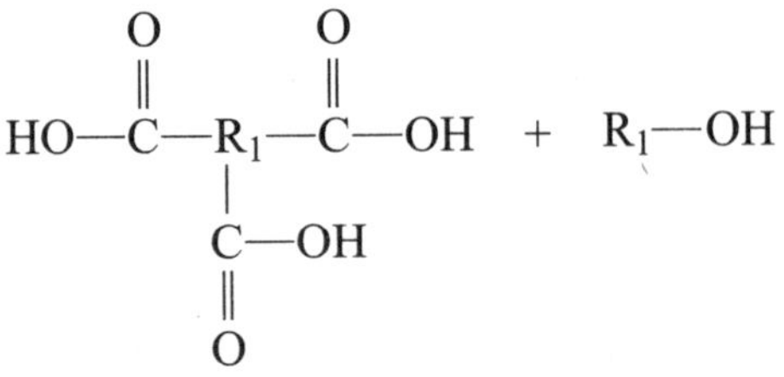

$$HO-\overset{\overset{\displaystyle O}{\|}}{C}-\underset{\underset{\underset{\underset{\displaystyle O}{\|}}{C-OH}}{|}}{R_1}-\overset{\overset{\displaystyle O}{\|}}{C}-OH \; + \; R_1-OH$$

3 moles of monohydric alcohol + 1 mole of triacid

$1 \times 3 + 3 \times 1 = 6$

$3 : 1 = 4$

average functionality = 6/4 = 1.5

In this reaction there will be no polymer formation because the average functionality of the system is less than two.

3.4 RELATIONSHIP BETWEEN AVERAGE FUNCTIONALITY, EXTENT OF REACTION AND DEGREE OF POLYMERIZATION (X_n)

W.H. Carothers derived a general expression, relating average functionality, extent of reaction and degree of polymerization and DP or $\overline{X}_n$ for polycondensation reactions carried out for a time period t. The expression is commonly known as Carother's equation. Let N_0 be the total number of reacting molecules initially present giving an average functionality 'f' for the system. The related functional groups are present in stoichiometric equivalence. Let N be the number of molecules present at time t when the extent of reaction is p. Then, number of molecules lost during the process over time period $t = (N_0 - N)$.

For each molecule lost, the number of functional groups lost is two (one of each kind) and hence the total number of functional groups lost is $2(N_0 – N)$ against the initial total number of $N_0 f$ functional groups.

Hence the extent of reaction $p = \dfrac{2(N_0 - N)}{N_0 f}$

or
$$p = \frac{2}{f}\left(1 - \frac{N}{N_0}\right) \tag{3.1}$$

The average degree of polymerization, $\overline{X}_n$ is defined as the average number of structural units per polymer molecule which can simply be expressed as

$$\overline{X}_n = \frac{N_0}{N} \tag{3.2}$$

Combining equations (3.1) and (3.2)

$$p = \frac{2}{f}\left(1 - \frac{1}{\overline{X}_n}\right) \tag{3.3}$$

Equation (3.3) is known as Carothers' equation and is very important for the understanding and control of the growth of polymers through poly-condensation reactions. The critical extent of reaction p_c at the gel point (for $\overline{X}_n \rightarrow c$) is given by

$$p_c = \frac{2}{f} \tag{3.4}$$

For a bifunctional system like A–A/B–B the value of p_c is 1 i.e., 100% monomer conversion. Thus the gelation starts occurring only after 100% monomer conversion which is a hypothetical situation. This is why polymers obtained in a bifunctional system are, under all practical situations, linear and hence soluble and fusible.

For a polyfunctional system, p_c can be calculated by the equation (3.4) and gel point may be predicted. For example, in a system having a functionality of 2.4 gelation starts after 83% conversion of monomers

$$p_c = \frac{2}{2.4} = 0.83 \text{ or } 83\%$$

3.5 MOLECULAR WEIGHT IN A STEP-GROWTH POLYMERIZATION

Polymer molecular-weight in a step-growth polymerization is determined by the fractional conversion or the extent of reaction, p, of the monomer during the polymerization.

$$\overline{X}_n = \frac{1}{1-p} \qquad \text{(Carothers' equation)} \qquad (3.5)$$

If the mean molecular weight of the repeat unit in polymer molecules is M_0, then the average molecular weight $\overline{M}_n$ of the polymer system is expressed as

$$\overline{M}_n = \overline{X}_n \cdot M_0 = \frac{M_0}{(1-p)} \qquad (3.6)$$

Similarly, the weight average degree of polymerization is given as

$$\overline{X}_W = \frac{1+p}{1-p} \qquad (3.7)$$

Small amounts of impurity and non-stoichiometric ratio of reactants would depress the molecular weight considerably.

If in the polymerization of a monomer of the type A–B, monofunctional impurity A is present and N_i is the number of molecules of A and N_0 is the number of molecules of A–B present then

$$\overline{X}_W = \frac{1+N_i/N_0}{1-p+N_i/N_0} \qquad (3.8)$$

If the polymerization reaction between two monomers A–A and B–B, which are present in the molar ratio r, is considered, the $\overline{X}_n$ will be given by the following expression.

$$\overline{X}_n = \frac{1+r}{2r(1-p)+1-r} \qquad (3.9)$$

3.6 KINETICS

It is similar to that of ordinary condensation reactions. A step growth polymerization may be first, second or third order depending upon whether it is an A–B or A–A/B–B and whether a catalyst is involved. The polymerization rate, R_0, may be expressed as the time-rate of change of monomer concentration. For a non-catalyzed A–A/B–B polymerization, this polymerization rate, as defined by the rate of disappearance of monomer, is second order in monomer concentration.

$$R_0 = -\frac{d\,[\mathrm{A-A}]}{dt} = k\,[\mathrm{A-A}]\,[\mathrm{B-B}] \tag{3.10}$$

$$k \rightarrow \text{rate constant}$$

Assuming a stoichiometric balance of monomer concentration Equation 3.10 can be simplified as

$$R_0 = -\frac{d\,[\mathrm{A-A}]}{dt} = k\,[\mathrm{A-A}]^2 \tag{3.11}$$

Integration of Equation 3.11 gives the relation

$$\frac{1}{[\mathrm{A-A}]} - \frac{1}{[\mathrm{A-A}]_0} = kt \tag{3.12}$$

$[\mathrm{A–A_0}]$ initial concentration at $t_0 = 0$.

The non-polymerized monomer concentration at any time t is related to the fractional conversion and initial monomer concentration by

$$[\mathrm{A–A}] = (1-p)\,[\mathrm{A–A}]_0 \tag{3.13}$$

Equation 3.13 can be rearranged for $(1-p)$

$$1 - \frac{[\mathrm{A-A}]}{[\mathrm{A-A}]_0} = P$$

or

$$\frac{[\mathrm{A-A}]_0 - [\mathrm{A-A}]}{[\mathrm{A-A}]_0} = \frac{[\mathrm{A-A}]_0 - kt}{1 + [\mathrm{A-A}]_0\,kt} \tag{3.14}$$

Combining Carotherts equation and Equation 3.14 gives

$$\overline{X}_n = 1 + [\mathrm{A-A}]_0\,kt$$

i.e., the number average chain length increases with time.

4

COPOLYMERIZATION

Learning Objectives

After studying this chapter, you will know:

- What are copolymers?
- What various types of copolymers can be prepared?
- What are chain-growth and step-growth copolymers?
- What are ionomers?
- What is the utility of copolymers?

Chapter Contents

• Copolymerization • Chain-growth copolymerization • Step-growth copolymerization • Ionomers • Polymer blends • IPNs • Utility of copolymers

4.1 CHAIN GROWTH COPOLYMERIZATION

The type of chain polymerization process in which two monomers are simultaneously polymerized is called chain copolymerization and the product is a chain growth copolymer.

4.1.1 Free Radical Copolymerization

In a free radical copolymerization involving two monomers, four separate propagation steps are possible. First monomer, M_1 may prefer to react with second monomer, M_2, or with itself and similarly second monomer may have its preference. Different types of copolymers are formed (alternate, random, block etc.) owing to the reactivity preference of the monomers.

$$M_1 + M_2 \longrightarrow M_1\, M_1$$
$$M_1 + M_2 \longrightarrow M_1\, M_2$$

$$\mathrm{M_1 + M_2 \longrightarrow M_2\,M_1}$$
$$\mathrm{M_1 + M_2 \longrightarrow M_2\,M_2}$$

Kinetics: Each propagation step has its own rate constant, i.e., (K_{ij}), where the first subscript (i), identifies the monomer at the end of the propagating chain prior to addition of monomer j.

$$\mathrm{\wedge\wedge\wedge\wedge\,M_1 + M_2 \longrightarrow M_1\,M_1}$$
$$\mathrm{\wedge\wedge\wedge\wedge\,M_1 + M_2 \longrightarrow M_1\,M_2}$$
$$\mathrm{\wedge\wedge\wedge\wedge\,M_1 + M_2 \longrightarrow M_2\,M_1}$$
$$\mathrm{\wedge\wedge\wedge\wedge\,M_1 + M_2 \longrightarrow M_2\,M_2}$$

The rates of disappearance of the monomer M_1 and M_2 can be obtained by considering the individual step, by which M_1 and M_2 are consumed. The rate equations are, therefore, given as

$$\frac{-d[\mathrm{M_1}]}{dt} = k_{11}\,[\wedge\wedge\wedge\mathrm{M_1}]\,[\mathrm{M_1}] + k_{21}\,[\wedge\wedge\wedge\mathrm{M_2}]\,[\mathrm{M_1}] \tag{4.1}$$

$$\frac{-d[\mathrm{M_2}]}{dt} = k_{12}\,[\wedge\wedge\wedge\mathrm{M_1}]\,[\mathrm{M_2}] + k_{22}\,[\wedge\wedge\wedge\mathrm{M_2}]\,[\mathrm{M_2}] \tag{4.2}$$

During a copolymerization, it is important to be able to predict how copolymer composition varies as a function of co-monomer reactivity and concentration at any time. The approach used relies on the realization that monomer consumed, during the copolymerization becomes incorporated into the copolymer structure and therefore, any relative change in the composition of the co-monomer mixture reflects the composition of copolymer formed at that instant of time. The relative change in the co-monomer composition is given by the instantaneous copolymerization equation, which is obtained by dividing Equation 4.1 by Equation 4.2.

$$\frac{-d[\mathrm{M_1}]}{-d[\mathrm{M_2}]} = \frac{[\mathrm{M_1}]\,k_{11}\,[\wedge\wedge\wedge\mathrm{M_1}] + k_{21}\,[\wedge\wedge\wedge\mathrm{M_2}][\mathrm{M_1}]}{[\mathrm{M_2}]\,k_{12}\,[\wedge\wedge\wedge\mathrm{M_1}] + k_{22}\,[\wedge\wedge\wedge\mathrm{M_2}][\mathrm{M_2}]} \tag{4.3}$$

Equation 4.3 can be expressed in a more concise and useful form by defining a monomer reactivity ratios. The reactivity ratios for monomer 1 and 2 are defined as

$$r_1 = \frac{k_{11}}{k_{12}} \tag{4.4 a}$$

$$r_1 = \frac{k_{22}}{k_{21}} \tag{4.4 b}$$

From the definitions of reactivity ratios (Equation 4.4(a) and (b)), it should be clear that when both reactivity ratios equal unity, there is no preferential monomer incorporation into the propagating chain, i.e.,

$$(k_{11} = k_{12}) \text{ and } (k_{22} = k_{21})$$

This means that the monomer sequence in the resulting copolymer is completely random. An example of random and ideal copolymerization is that of styrene and 4-chloro styrene.

When both reactivity ratios are zero, i.e., $k_{11} = k_{22} = 0$, the monomer sequence will be perfectly alternating, e.g., copolymerization of styrene and maleic anhydride. If both reactivity ratios are small, but not exactly zero, the co-monomer sequence will be highly alternating with some random placement as in the case of copolymers of styrene and acrylonitrile.

When both reactivity ratios are very much larger than unity (i.e., $K_{ii} >> K_{ij}$) only homopolymers or block copolymers will be formed.

The reactivity ratios for monomer pairs can be calculated by means of the Q-e-Scheme. In this approach, the propagation rate constant (K_{ij}) is defined as

$$K_{ij} = P_i Q_j \exp(-e_i e_j) \tag{4.5}$$

where, P_i is proportionality constant, Q_j is measure of monomer reactivity, e = polarity of the radical M_1 or M_2.

It then follows from the definition of the reactivity ratios that

$$r_2 = \frac{K_{22}}{K_{21}} = \left(\frac{Q_2}{Q_1}\right) \exp\left[-e_2(e_2 - e_1)\right] \tag{4.6a}$$

and

$$r_2 = \frac{K_{22}}{K_{21}} = \left(\frac{Q_2}{Q_1}\right) \exp\left[-e_2(e_2 - e_1)\right] \tag{4.6b}$$

In general, the value of a reactivity ratio is independent of the nature of the initiator and solvent in a free radical copolymerization, however, there is a weak temperature dependence.

A limitation of the copolymerization (Equation 4.3) is the use of radical chain concentration, which are usually not known and easily measured. By using Equation (4.4a, b) and assuming a steady state radical concentration for M_1 and M_2 during the propagation the instantaneous copolymerization equation becomes

$$\frac{d[M_1]}{d[M_2]} = \frac{[M_1]}{[M_2]} = \frac{r_1[M_1]+[M_2]}{[M_1]+r_2[M_2]} \tag{4.7}$$

From this equation, copolymer composition can be estimated if reactivity ratios are known or values for the reactivity ratios can be obtained, if the monomer concentrations and copolymer compositions are known.

The copolymerization equation 4.7 may be recast in an alternative form, i.e., useful for calculating the instantaneous copolymer composition for a given monomer concentration. For this purpose, the mole fraction of M_1 in the reaction mixture is defined as —

$$f_1 = \frac{[M_1]}{[M_1]+[M_2]} \tag{4.8}$$

where, $f_1 + f_2 = 1$

In a similar fashion, the mole fraction of monomer 1, F_1 in the copolymer is given by the reactive differential change in monomer concentration as

$$F_1 = \frac{d[M_1]}{d[M_1]+d[M_2]} \tag{4.9}$$

where, $F_1 + F_2 = 1$.

If the numerator and denominator of the Equation 4.9 are each divided by $d[M_2]$, use of Equations 4.7 and 4.8 gives the following useful relationship.

$$F_1 = \frac{[M_1]\left(\frac{r_1[M_1]+[M_2]}{[M_1]+r_2[M_2]}\right)}{[M_1]\left(\frac{r_1[M_1]+[M_2]}{[M_1]+r_2[M_2]}+1\right)}$$

$$\frac{M_1(r_1[M_1]+[M_2])}{M_1(r_1[M_1]+[M_2]+[M_1]+r[M_2]}$$

$$F_1 = \left(\frac{r_1 f_1^2 + f_1 f_2}{r_1 f_1^2 + 2 f_1 f_2 - r_2 f_2^2}\right) \tag{4.10}$$

4.2 STEP GROWTH COPOLYMERIZATION

Step polymerization is a versatile means of synthesizing a variety of different polymers. It is very much possible to prepare polymers that can have two types of repeating units in their structures by step growth polymerization. These polymers are referred to as step growth copolymers.

4.3 IONOMERS

An ionomer, as one might guess from the name, is an ion containing polymer. But an ionomer is more than just a polymer with ionic groups. Any polymer with ionic groups is called a polyelectrolyte. But an ionomer is a special kind of polyelectrolyte. First of all, they are copolymers. They contain both nonionic repeat units, and a small amount of ion containing repeat units. The ionic groups make up less than 15% of the polymer.

One example of an ionomer is polyethylene-co-methacrylic acid. This polymer is a sodium or zinc salt (which provides the ions) of copolymers derived from ethylene and methacrylic acid.

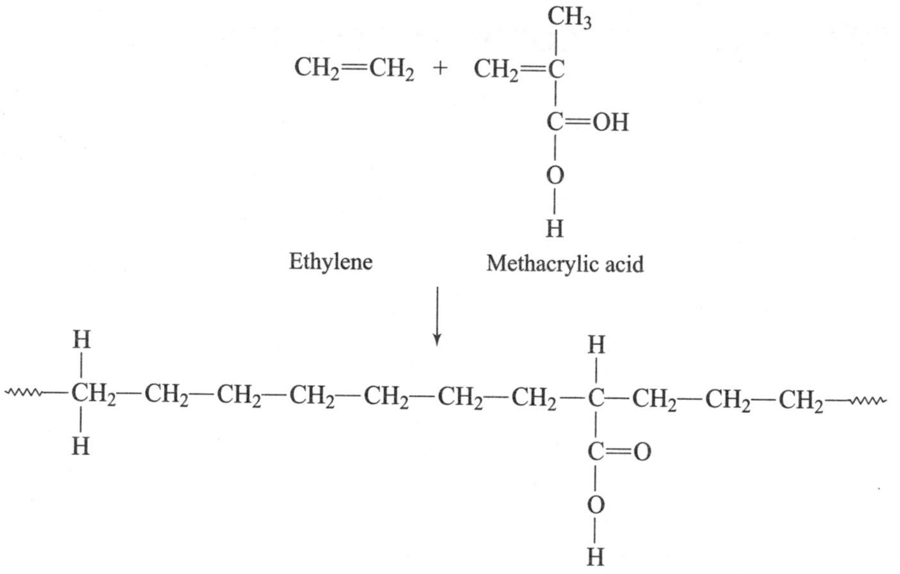

$$CH_2{=}CH_2 + CH_2{=}C(CH_3){-}C({=}OH){-}O{-}H$$

Ethylene Methacrylic acid

$$\downarrow$$

$$\sim\!\!\sim CH_2{-}CH_2{-}CH_2{-}CH_2{-}CH_2{-}CH_2{-}CH_2{-}C(H)(C{=}O{-}O{-}H){-}CH_2{-}CH_2{-}CH_2\sim\!\!\sim$$

Poly(ethylene-co-methacrylic acid)

If the acidic groups are neutralized by NaOH, a sodium salt of copolymer is formed.

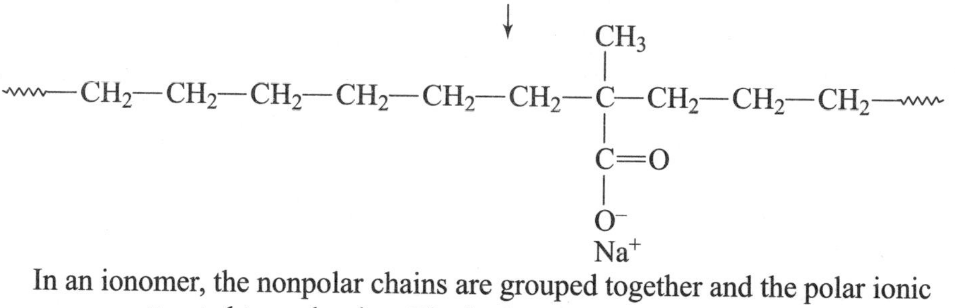

$$\downarrow$$

$$\sim\!\!\sim CH_2{-}CH_2{-}CH_2{-}CH_2{-}CH_2{-}CH_2{-}C(CH_3)(C{=}O{-}O^-\,Na^+){-}CH_2{-}CH_2{-}CH_2\sim\!\!\sim$$

In an ionomer, the nonpolar chains are grouped together and the polar ionic groups are attracted to each other. The ionic groups would like to go off into a little corner by themselves, but since they are attached to the polymer chain, they cannot. This allows thermoplastic ionomers to act in ways similar to that of crosslinked polymers or block copolymers.

4.4 POLYMER BLENDS AND INTERPENETRATING POLYMER NETWORKS

Polymer blends and interpenetrating polymer network (IPN) are different from copolymers but like copolymers are used to bring together the properties of different polymers. The total of all polymer blends (produced by both step and chain reaction) is estimated at about 3% of the total polymer production—about 2 billion pound per year in the United States. There is a considerable activity in this area since "new" products can be obtained and markets expand by the physical mixing together of existing products and no new polymer needs to be synthesized.

4.5 UTILITY OF COPOLYMERIZATION

There has been enormous commercial success in the synthesis of various copolymers. Intense research activity continues in this area of polymer science with enormous number of different combinations of repeating units being studied. Copolymer synthesis offers the ability to alter the properties of a homopolymer in desired directon by the introduction of an appropriately chosen second repeating unit. One has the potential to combine the desirable properties of two homopolymer into a single copolymer. Copolymerization is used to alter such properties as crystallinity, flexibility, T_m, T_g. The magnitudes and sometimes even the directions, of the property alteration differ depending on whether statistical alternating or block copolymer is involved. The crystallinity of statistical copolymer is lower than that of either of the respective homopolymers (i.e., the homopolymers corresponding to the two different units) because of the decrease in structural regularity. The melting temperature of any crystalline material formed is usually lower than that of either homopolymer. The T_g value will be in between those for the two homopolymers.

Alternating copolymers have a regular structure, and their crystallinty may not to be significantly affected unless one of the repeating units contain rigid, bulky, or excessively flexible chain segments. The T_m and T_g value of an alternating copolymer are in between the corresponding values for the homopolymers. Block copolymers shows the behavior (crystallinity, T_m, T_g) present in the corresponding homopolymer as long the block lengths are not too short. This behavior is typical since A block from different polymer molecules aggregate with each other and separately, B block from different polymer molecules aggregate with each other. This offers the ability to combine the properties of two very different polymers into one block copolymer. The exception to this behavior occurs infrequently when the tendency for cross aggregation between A and B blocks with B block.

Most commercial utilization of copolymerization fall into one of the two groups. One group consist of various statistical (usually random) copolymers in which the two repeating units possess the same functional groups. The other groups of commercial copolymers consist of block copolymers in which two repeating units have different functional groups. There are very few commercial statistical copolymers in which the two repeating units have different functional groups. The reason for this situation probably resides in the difficult finding one set of reaction condition for simultaneously performing two different reactions.

5

POLYMER PROPERTIES

Learning Objectives

After studying this chapter, you will know:

- What is structure property relationship in polymers?
- What is the significance of various types of average molecular weights?
- What is polydispersity index?
- What are the solution properties of polymers?
- About deformation in polymers and how are they measured?
- About flow behaviour of polymer melts and solutions.
- What types of transitions do occur in polymer molecules?
- How glass transition temperature is measured?
- About degradation of polymers.

Chapter Contents

• Structure property relationship • Average molecular weights • Polydispersity index • Solution properties of polymers • Deformation in polymers • Flow behavior of polymer melts and solutions • Degradation of polymers

The wide variety of natural, semisynthetic and synthetic polymers, known today, exhibit wide diversity of properties. Some are rigid, hard and strong and dimensionally stable, while others are soft, flexible or largely deformable under stress. Some are soluble and fusible while others are more resistant to heat and solvents and may be even insoluble and infusible.

All such properties vary from a polymer of one type to a polymer of another type. They may vary even between samples of the same type of polymer depending on how they were prepared and treated thermomechanically before being tested. Consideration of chemical structure of repeat unit to predict properties has certain limitation, particularly the mechanical behavior. Chemical structure contributes partly and often indirectly to mechanical properties. More consequential and direct role is played by the supramolecular structure, i.e., the physical arrangement of the chain molecules with respect to each other.

5.1 STRUCTURE-PROPERTY RELATIONSHIP

Polymers are long, complicated objects; they might be equally complicated to deal with. But for many properties this is not true. Many of their properties depend much more strongly on their length (or equivalently degree of polymerization n), than on specific chemistry. Also a single chain tends to interact with many other chains. This means we can average out the interactions of surrounding chains and use so called mean field theory and statistical mechanics to understand their behavior. This leads to so called scaling laws. It turns out that mean field theories are better obeyed by polymers than for many simpler, atomic species.

There are several factors other than the length of polymer chain which determine the physical properties of a polymer; the secondary valency forces, polymer configuration and conformation, and the nature of chain packing of molecules. These factors basically influence the crystallinity of polymers, which in turn, along with other factors influence mechanical, thermal, solution and rheological properties.

5.2 AVERAGE MOLECULAR WEIGHT

For any chemical substance, the molecular weight is calculated by the following equation.

$$M = \frac{W}{N} \quad \text{Molecular Weight} = \frac{\text{Sample weight}}{\text{No. of moles in sample}}$$

Unlike a normal chemical substance, a polymer is usually a complex mixture of molecules of different molecular weights. The polymers are thus polydisperse and heterogeneous in nature i.e. all polymer molecules in a polymer sample are not of equal size, they vary in size and thus their molecular weights are also different. Where a distribution of molecular weight exists, an average molecular weight is calculated. Therefore, the molecular weight of a polymer sample is actually an average of the molecular weights of constituents molecules. Different averages are obtained depending upon the method of determination of molecular weight. The common methods for averaging are:

(i) Number average molecular weight: The number average molecular weight M_n is defined as-

$$\overline{M_n} = \frac{W}{N} = \frac{\sum_{i=1}^{\infty} n_i M_i}{\sum_{i=1}^{\infty} n_i} \tag{5.1}$$

$$= \frac{n_i M_i}{\sum_{i=1}^{\infty} n_i} + \frac{n_2 M_2}{\sum_{i=1}^{\infty} n_i} + \cdots$$

$$= \sum_{i=1}^{\infty} \left(\frac{n_i}{N}\right) M_i$$

W = total sample weight

W_i = total weight of i-mer

N = total number of moles in sample

n_i = number of moles of i-mer

M_i = molecular weight of i-mer

n_i/N = mole fraction of i-mer

Suppose we have a polymer sample and it has polymer molecules of five different sizes. Molecular weight of this sample is calculated in this way.

Total number of molecules $n = n_1 + n_2 + n_3 + n_4 + \ldots\ldots\ldots + n_i = \Sigma n_i$

Number of molecules in fraction 1 = $\dfrac{n_1}{\Sigma n_i}$

Molecular weight contribution by fraction 1 = $\dfrac{n_1 M_1}{\Sigma n_i}$

Similarly, molecular weight contribution by other fraction will be as follows:

$$\frac{n_2 M_2}{\Sigma n_i}, \frac{n_3 M_3}{\Sigma n_i} \cdots \frac{n_i M_i}{\Sigma n_i}$$

$$\overline{M_n} = \frac{n_1 M_1}{\Sigma n_i} + \frac{n_2 M_2}{\Sigma n_i} + \ldots + \frac{n_i M_i}{\Sigma n_i}$$

$$\overline{M_n} = \frac{n_i M_i}{\Sigma n_i}$$

(ii) Weight average molecular weight

Total weight of polymer sample = $W = \Sigma n_i M_i$

Weight fraction of fraction 1 = $\dfrac{n_1 M_1}{\Sigma n_i M_i}$

Molecular weight contribution by fraction 1 = $\dfrac{n_1 M_1 M_1}{\Sigma n_i M_i} = \dfrac{n_1 M_1^{\,2}}{\Sigma n_i M_i}$

$$\overline{MW} = \frac{n_1 M_1^2}{\Sigma n_i M_i} + \frac{n_2 M_2^2}{\Sigma n_i M_i} + \ldots + \frac{n_i M_i^2}{\Sigma n_i M_i}$$

$$M_w = \frac{n_1 M_1^2}{\Sigma n_i M_i} \tag{5.2}$$

(iii) Viscosity average molecular weight

Before coming to viscosity average molecular weight, one should know the definitions of the terms given below:

Relative viscosity (η_{rel}) is the ratio of the dynamic viscosity of the solution to that of the pure solvent ($\eta_{rel} = \eta/\eta_0$ where η is the dynamic viscosity of the solution and η_0 is the dynamic viscosity of the solvent). As it is a ratio, it is dimensionless with no units. It is related to the intrinsic viscosity $[\eta]$ by the Huggins-Kramer equation: In $(\eta_{rel})/c = [\eta] + k_2[\eta]^2 c$, where c is the concentration.

η = viscosity of solution
η_s = solvent viscosity
$\eta_r = \eta/\eta_s = t/t_0$ (relative viscosity)

Specific viscosity (η_{sp}) is one less than the relative viscosity ($\eta_{sp} = \eta_{rel} - 1$; $\eta_{sp} = (\eta - \eta_0)/\eta_0$ where η is the dynamic viscosity of the solution and η_0 is the dynamic viscosity of the solvent). As with the relative viscosity, it has no units.

$$\eta_{sp} = \frac{\eta - \eta_s}{\eta_s} = \eta_r - 1$$

Reduced viscosity (η_{red}) is the specific viscosity divided by the concentration. It has units of reciprocal concentration, e.g. ml g^{-1}. It is related to the intrinsic viscosity $[\eta]$ by the Huggins-Kramer equation: $\eta_{red} = [\eta] + k_1[\eta]^2 c$, where c is the concentration.

$\eta_{red} = \eta_{sp}/c$ (reduced viscosity)
$\eta_{inh} = \log \eta_r/c$ (inherent viscosity)
c = concentration of polymer

$$[\eta_{int}] = \lim_{c\to 0} \eta_{red} = \lim_{c\to 0} \eta_{inh}$$

This intercept is

$$\frac{[\eta]}{M_v} = \left(\frac{[\eta]}{K}\right)^{1/a}$$

$$M_v = \frac{n_1 M_1^2}{\Sigma n_i M_i} \tag{5.3}$$

This equation is applicable when molecular weight distribution in a series of polymer sample does not differ too much.

(iv) Z-Average molecular weight (z stands for centrifuge)

$$M_z = \frac{\Sigma N_i M_i^3}{\Sigma N_i M_i^2} \tag{5.4}$$

5.2.1 Polydispersity Index

The polydispersity index (PDI), is a measure of the distribution of molecular weights in a given polymer sample. The PDI calculated is the weight average molecular weight divided by the number average molecular weight. It indicates the distribution of individual molecular weights in a batch of polymers. The PDI has a value always greater than 1, but as the polymer chains approach uniform chain length, the PDI approaches unity.

The ratios of different molecular weight averages M_w/M_n and M_z/M_w are often used as characteristics of polymer polydispersity. Mathematically, $M_n = M_w = M_z$; here the equality sign holds only for monodisperse polymers.

For most polymers M_w/M_n: 1.5–2.5

For monodisperse polymer: $M_w/M_n = M_z/M_w = 1$

The following curve shows a distribution of polymer molecular weights which illustrates that the weight average molecular weight is higher than the number average molecular weight.

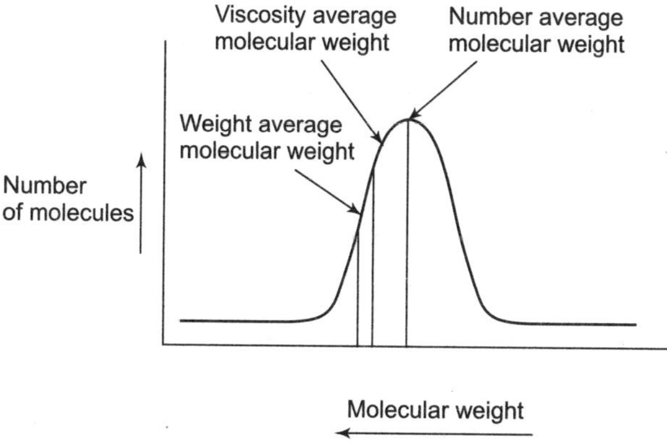

Fig. 5.1 Molecular Weight Distribution

5.2.2 Effect of Polymerization Mechanism on PDI

Typical PDIs vary based on the mechanism of polymerization and can be affected by a variety of reaction conditions. In synthetic polymers, it can vary greatly due to the reactant ratio, how close the polymerization went to completion etc. For typical addition polymerization, values of the PDI can range around 10 to 20. For typical step polymerization, values range around 2 to 3—Carother's equation limits PDI to values of 2 and above. Living polymerization, a special case of addition polymerization, leads to values very close to 1. Such is the case also in biological polymers, where the PDI can be very close or equal to 1, indicating only one length of polymer which is present.

5.3 POLYMER SOLUTIONS AND SOLUBILITY

Solvents are frequently used during the polymerization process, during fabrications and for determination of molecular weight. The thermodynamics and statistics of polymer solutions is an important branch of physical chemistry. The topics of practical importance are: general rules for polymer solubility, thermodynamic basis of polymer solubility, solubility parameters, polymer melts.

5.3.1 Dissolution of Polymers

The dissolution of a polymer is generally a slow process, which can take even several weeks, depending on the structure and the molecular weight of a given polymer.

(a) Polymer molecules in solid state just after being added to a solvent

(b) **First step:** a swollen gel in solvent

(c) **Second step:** solvated polymer molecules dispersed into a solution

Fig. 5.2 Schematic representation of the dissolution process for polymer molecules

5.3.2 General Rules for Polymer Solubility

These are some general experimental observations on the dissolution of polymers.

(i) Like dissolves like, i.e., polar solvents will tend to dissolve polar molecules and non-polar solvents to dissolve non-polar molecules. Chemical similarity of polymer and solvent is a fair indication of polymer solubility, e.g. polyvinylalcohol dissolves in water as both are polar and polystyrene dissolves in toluene as both are non-polar.

(ii) In a given solvent, at a particular temperature, the solubility of a polymer will decrease with the increasing molecular weight.

(iii) (a) Cross-linking eliminates solubility.

(b) Crystallinity, in general, acts like cross-linking, but it is possible in some cases to find solvents strong enough to break the crystalline bonding forces and dissolve the polymer. Heating the polymer towards its crystalline melting point allows its solubility in appropriate solvents.

(iv) The rate of polymer solubility—
   (a) increases with short branches, which loosen up the main chain structure, allowing the solvent molecules to penetrate more easily.
   (b) decreases with longer branches, because the entanglement of these branches makes it harder for the individual molecules to separate.
   (c) decreases with the increasing molecular weight.

Rules (i), (ii) and (iii) are equilibrium phenomenon and therefore describable thermodynamically, while rule (iv) is a rate phenomenon and is governed by rates of diffusion of polymer in solvent.

5.3.3 Thermodynamic Basis of Polymer-Solubility

$\Delta G = \Delta H - T.\Delta S$

Only if ΔG is negative, will the solution process be thermodynamically feasible. T and ΔS should be positive, because in a solution, the molecules are in more random state, than in solvent. The change in enthalpy may be either positive or negative. A positive ΔH means polymer and solvent prefer their own company, i.e., they are in low energy state while negative ΔH means, the solution is in lower energy state. Minus ΔH occurs when there is the formation of secondary bonds between polymer and solvent. If ΔH is positive, its value should be lower than the value of T. ΔS.

The entropy change in forming a solution of polymer is quite small. The reason for this can be explained by a two-dimensional lattice model.

In the low molecular weight mixture, the solute molecules may be distributed randomly throughout the lattice, the only restriction being that a lattice side cannot be occupied, simultaneously by two or more molecules. This gives rise to a large number of configurational possibilities, i.e., a high entropy.

• → Polymer molecule

O → Solvent molecule

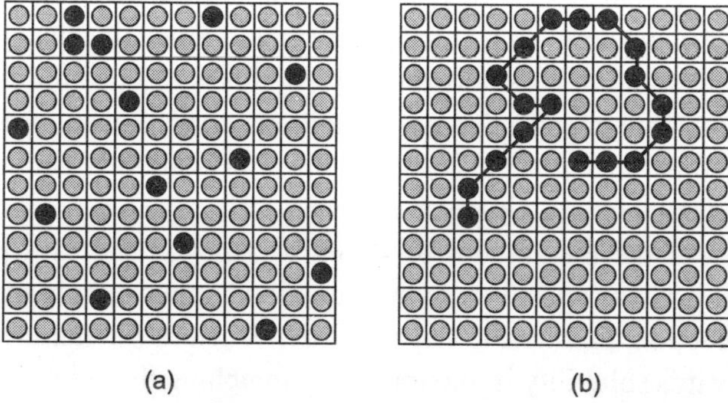

Fig. 5.3 Two-dimensional lattice model of solubility for (a) a low molecular weight solute (b) for a polymer solute

In the polymer solution, however, each chain segment is confined to the lattice-site adjacent to the next chain segment; greatly reducing the configurational possibilities. For a given number of chain segment (equivalent masses or volumes of polymer), the more chain, they are split up into (i.e., the lower the molecular weight), the higher will be their entropy upon solution. This explains directly the second rule, the decrease in solubility with the increasing molecular weight. In general, for high molecular weight polymers, because the $T.\Delta S$ term is so small, if ΔH is positive, it must be even smaller (nearly zero), if the polymer is to be soluble.

5.3.4 Solubility Parameters

For regular solutions (those in which solute and solvent do not form specific interaction), the change in the internal energy upon solution is given by:

$$\Delta H \cong \Delta E = v_1 v_2 (\delta_1 - \delta_2)^2 \left(\frac{\text{Cal}}{\text{cm}^3 \text{ solution}} \right) \tag{5.5}$$

where, ΔE is the change in internal energy in the solution process, v is the volume fraction, and δs are solubility parameters. The subscripts 1 and 2 usually refer to solute (polymer) and solvent respectively. The solubility parameter is defined as follows:

$$\delta = (\text{CED})^{1/2} = \left(\frac{\Delta E_v}{v} \right)^{1/2} = \left(\frac{\text{Cal}}{\text{cm}^3} \right)^{1/2} \tag{5.6}$$

where CED is Cohesive Energy Density, [a measure of the strength of the intermolecular forces holding the molecules together in the liquid state]; ΔE_v is molar change in internal energy on vaporization (unit Cal g/mol), ΔH is latent heat of vaporization, and v is the molar volume for liquid (cm^3/g mol).

$$\delta = \left(\frac{\Delta H - RT}{v} \right)^{1/2} \tag{5.7}$$

For a process that occurs at constant volume and constant pressure, the changes in internal energy and enthalpy are equal. Since the change in volume from solvent to solution is usually quite small, this is a good approximation for the dissolution of polymers under most conditions, so the Equation (5.1) provides a means of estimating enthalpies of solutions if the solubility parameters of the solvent and parameters are known.

Regardless of the magnitudes of δ_1 and δ_2, they are always positive, the predicted ΔH is always positive, so this equation applies only in the absence of specific interaction, that lead to negative ΔH. When ΔH is minimized and the tendency towards solubility is maximized by matching the solubility parameter as closely as possible.

As a rough rule of thumb, difference between δ_1 and δ_2 should be less than 0.5 for polymer solubility, but this applies only in the absence of specific interaction. Measuring the solubility parameter of a low molecular weight solvent is a problem. Polymers, on the other hand, degrade long before reaching their vaporization temperatures, making it impossible to evaluate ΔE_v directly. Fortunately, there is a way around this difficulty. The greatest tendency of a polymer to dissolve, occurs when its solubility parameter matches with that of a solvent.

If the polymer is cross-linked lightly, it cannot be dissolved, but only swells. The maximum swelling will be deserved when the polymer and solvent solubility parameters are matched. So polymer-solubility parameters are determined by soaking lightly cross-linked samples in a series of solvents of known solubility parameters. The value of the solvents at which maximum swelling is observed is taken as solubility parameter of the polymer.

Solubility parameters of solvent mixture can be calculated by

$$\delta_{\text{mixture}} = \frac{(X_1V_1\delta_1 + X_2V_2\delta_2)}{(X_1V_1 + X_2V_2)} \tag{5.8}$$

where X = mole fraction

$$\delta_1 - \delta_2 \leq 0.5 \tag{5.9}$$

While the solubility parameter concept has proved useful, there are unfortunately many exceptions to the above equation. Polymer solubility is too complex a phenomenon to be described quantitatively with a single parameter. There are several other engineering models on the basis of those solubility of a polymer sample can be predicted.

Hansen's Three-dimensional Solubility parameter model

According to this, the total change in internal energy on vaporization (i.e., ΔE_v), may be considered the sum of three individual contributions, one due to hydrogen bonds and another due to permanent dipole interactions and a third from the dispersion forces.

$$\Delta E_v = \Delta E_h + \Delta E_p + \Delta E_d \tag{5.10}$$

divided by the molar volume v, gives

$$\frac{\Delta E_v}{v} = \left(\frac{\Delta E_h}{v}\right) + \left(\frac{\Delta E_p}{v}\right) + \left(\frac{\Delta E_d}{v}\right) \tag{5.11}$$

or

$$\delta^2 = \delta_h^2 + \delta_p^2 + \delta_d^2 \tag{5.12}$$

The solubility parameter (δ) may be thought of as a vector in a three-dimensional (δ_p, δ_d, δ_h) space.

Equation (5.12) gives the magnitude of the vector in terms of its components. A solvent or polymer therefore with given values of δ_p, δ_d and δ_h is represented as a point in space with δ being the vector from the origin to this point.

Furthermore, it has been found on a purely impirical basis that if δ_d is plotted on a scale twice the size as that used for δ_p and δ_h, all solvents that will dissolve that polymer fall within a sphere of radius R, surrounding the points δ_p, δ_d, δ_h for the polymer.

Despite its impirical nature, the three-dimensional solubility parameter has proved of greater practical utility, particularly in the paint industry, where the economic choice of solvent is of critical importance. It is capable of explaining those cases in which solvent and polymer solubility parameters are almost perfectly matched and yet the polymer will not dissolve (the δ vectors have the same magnitude but different directions) or where two non-solvents can be mixed to form a solvent (the solvent components lie on opposite sites outside the sphere, the mixture within). Inorganic pigments may also be characterized by δ vectors. Pigments, where δ vectors closely match with those of a solvent tend to form stable suspension in that solvent.

5.3.5 Properties of Dilute Solutions

In a fairly dilute solution of polymer, there are not too many entanglement between the molecules. In a good solvent (one whose solubility parameter closely matches with that of polymer), the secondary forces between the polymer segments and solvent molecules are strong and the polymer molecule will assume spread out conformation in solution. In a poor solution, the attractive forces between the segments of polymer chains are greater than those between the chain segment and the solvents, i.e., the chain segments prefer their own company and the chain will bolt-up tightly.

Polymer in ball-up condition

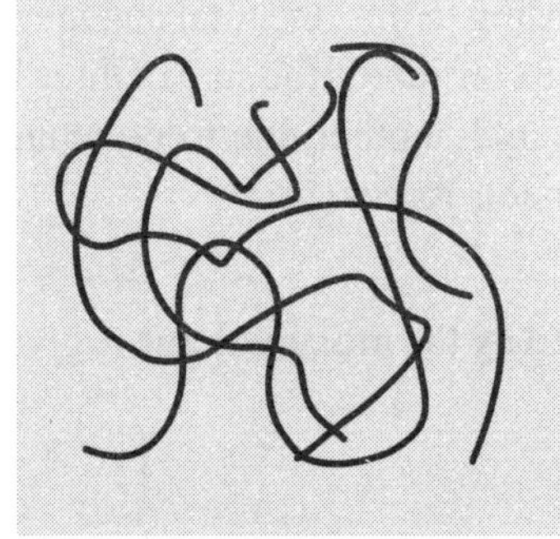

Polymer in spread-up condition

Fig. 5.4 Effect of Solvent and Temperature on a Polymer

Imagine a polymer in a good solvent and a non-solvent is now added. Ultimately a point is reached where the mixed solvent becomes too poor to sustain solution and polymer precipitates out, at some point when $\Delta G = 0$ and $\Delta H = T.\Delta S$. This point obviously depends on the temperature, polymer molecular weight (mainly through its influence on ΔS) and polymer solvent interaction (mainly through its influence on ΔH).

Adjusting either the temperature or the polymer solvent system, allows fractionation of polymer according to molecular weight as successively smaller molecules precipitate upon lowering the temperature or going to poorer solvent. In the limit of infinite molecular weight (the minimum possible ΔS), the situation in which $\Delta H = T.\Delta S$, is known as θ-condition. Under these conditions the polymer solvents and polymer-polymer interactions are equal and the solution behaves in a so called ideal fashion with the record virial coefficient equal to zero and so on.

For a given polymer, the θ conditions can be reached at a fixed temperature by adjusting the solvent to give a θ solvent or with a particular solvent by adjusting the temperature to reach the θ of the flory temperature. Any actual polymer will be soluble under θ conditions, because of its lower than infinite molecular weight and consequently larger ΔS.

Influence of Solvent on Viscosity

Going from good to poor solvent the polymer molecules ball up, lowering the solution viscosity (when polymer concentration is kept constant). Thus solution viscosity can be controlled by adjusting solvent. This fact is of good importance to the surface coating industry. For example, in formulating a lacquer (a solution of polymer in solvent + pigment which dries only by solvent evaporation), the viscosity for optimum spraying or brushing characteristics might be obtained with a mixed solvent with the poorer component more volatile. After application, the poorer components evaporate first leaving a high viscosity and therefore sag and run resistance film on the substrate.

Concentrated Solutions (Plasticizers)

A pure polymer (amorphous) consists of entangled mass of polymer chains. The ease with which this mass can deform depends on the ability of the polymer chains to entangle and slip past one another.

One way of increasing this ability is to raise the temperature, another is to add a low molecular weight liquid an external plasticizer to the polymer. By forming secondary bonds to polymer molecules and spreading them apart, the plasticizer reduces the polymer-polymer secondary bonding and provides more room for the polymer molecules to move around giving a softer, more easily deformable mass.

5.4 SOLID STATE PROPERTIES

Polymer Morphology

The study of physical form, arrangement and structure of the molecules of a material comes under the word, morphology. Polymer morphology covers the study of the arrangement of macromolecules into amorphous and crystalline regions and the structure of the molecular aggregate in totality.

Some commercial polymers particularly the polyolefins, a highly crystalline material with well-defined crystalline morphology (outer structure's study) consisting of chain-folded lamella joined in supramolecular (physical) structure, called spherulites. Although single crystals of some polymers such as polyethylene can be grown under laboratory conditions, no bulk polymer is completely crystalline. In the case of semi-crystalline polymers, regular crystalline units are linked by un-oriented random conformation chains that constitute amorphous region. Other polymers may have very low crystallinity characterized by poorly defined crystalline microstructure in an amorphous matrix. An important example of commercial polymer with low crystallinity is polyvinyl chloride (P.V.C.). Many other polymers such as atactic PS (Polystyrene) and PMAA (polymethyl methacrylate) are totally amorphous. In all these cases, morphological aspects such as the presence of crystalline structure have a significant influence on the thermal, physical and mechanical properties of the polymer.

5.4.1 Amorphous State

Amorphous polymers are completely devoid of any molecular order. Completely amorphous polymer like atactic polystyrene exists as long, randomly coiled, interpenetrating chains that are capable of forming stable, flow-restricting entanglements at sufficiently high molecular weights.

5.4.2 Crystalline State

5.4.2.1 Requirements for Crystallinity

(i) An ordered regular chain structure is necessary to allow the chains to pack into a regular crystal lattice. Thus stereo-regular polymers are more likely to be crystalline than those that have irregular chain structures. Irregular protruding side gaps, interfere with the arrangements of the main chains in a regular lattice and hinder crystallinity.

(ii) No matter how regular the chain is, the secondary forces holding the chains together in the crystal lattice must be strong enough to overcome the disordering effect of thermal energy, so hydrogen bonding or strong dipole interaction promotes crystallinity and other things being equal, raise the crystalline melting point.

Ordering of Polymer Chain

There are several models available to explain the ordering of chains or molecular order in crystalline polymers:

(i) Fringed miscelle model
(ii) Folded chain crystallites
(iii) Extended chain crystals
(iv) Spherulites

Under favorable conditions, some polymers cooled from the melt can organize into regular crystalline structure. Such crystalline polymers have less perfect organization than the crystals of low molecular weight compounds. The basic units of crystalline polymer morphology are crystalline lamellae. Lamellae consist of arrays of folded chains and re-entry of each chain in the folded structure can be adjacent or non-adjacent. A chain participating in adjacent re-entry can form a tight (or regular) loop or form a loose loop (irregular). The thickness of typical crystallite may be 100–200 Å indicating that only a portion of the complete chain is included in each fold.

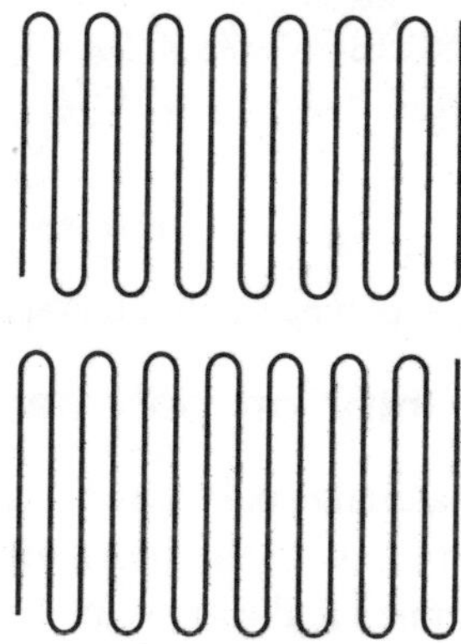

Fig. 5.5 The Ideal Lamellar Stacking

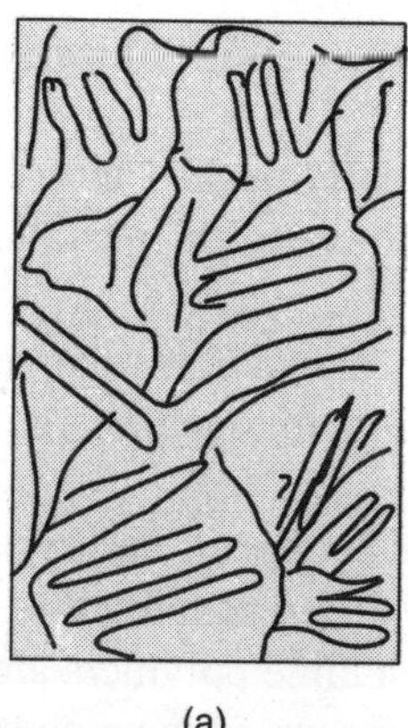

(a)

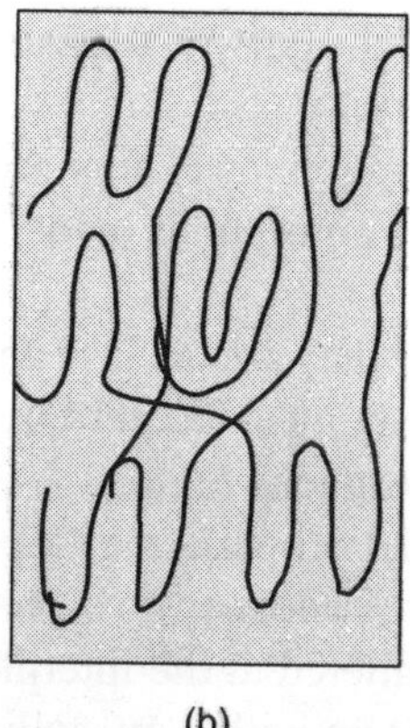

(b)

Fig. 5.6 (a) Fringed Miscelle Model and (b) Interlamellar Amorphous Model

The ordering of chains showing stacks of lamellae interspaced with and connected by amorphous regions is referred to as the interlamellar amorphous model. In fringed miscelle model, amorphous material is interspaced between the randomly distributed crystallites. High thermal energy favors a large number of conformations in the melt state. As the melt is cooled, the low energy conformations are favored and in most of the polymers, the chains organize in the extended chain conformation. In case of polymers containing larger pendant groups, the lowest energy conformation is a helix of some preferred geometry.

For some polymers during crystallization, ordering of chains can be in the form of large spherical structures called spherulites.

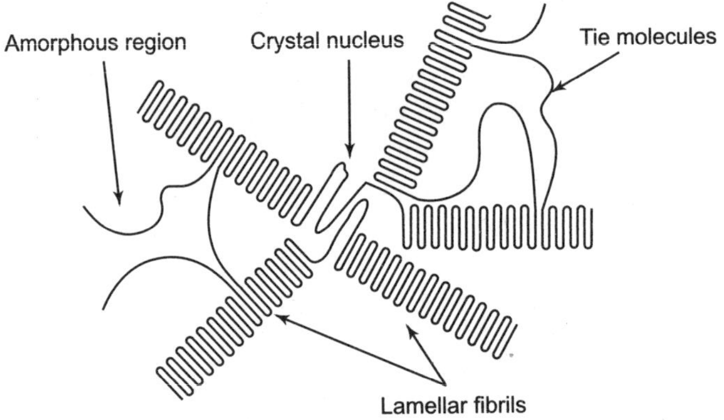

Fig. 5.7 A polymer crystalline spherulites

The effect of crystallinity on polymer property

The presence of crystalline material in polymer strongly influences their property. This influence is most profound on the mechanical property. Crystallinity significantly increases the strains and rigidity of a polymer because—

(i) Polymer chains are packed together more efficiently, closely and tightly in crystalline regions, than in amorphous areas. Therefore, crystallites will have a high density (density is in fact a convenient measure of the crystallinity).

(ii) Since the polymer chains are more tightly packed in crystalline region than in the amorphous areas, there are more of them available to support a stress.

(iii) Also since they are in close and regular contact over relatively low density in the crystallites, the net force holding them together are far greater than in the amorphous regions.

(iv) The optical properties of a polymer also depend on crystallinity. When light passes between two phases with different refractive indexes, some of it is scattered at the interphase. Crystalline polymers are actually two-phase system with a crystalline phase dispersed in an amorphous matter. The denser crystalline areas have a higher refractive index than the amorphous areas, so the crystalline polymers are either opaque or translucent. The dimensions of the crystallites become appreciably smaller than the wavelength of light, the amount of light scattered decreases. And the polymer with very small crystallites and a low degree of crystallinity might appear transparent. So in general transparent polymers are completely amorphous. The lack of transparency in pure homopolymer sample, is a sure sign of crystallinity, but the translucency or opaqueness can be created by fillers.

5.5 DEFORMATIONS IN POLYMERS

Deformations are primarily of two kinds:

(i) Ideal elastic – which takes place instantaneously due to the effective force by applied stress and disappears upon release of the same (temporary and reversible deformation).

(ii) Flow – characterized by permanent and irreversible deformation.

Under processing conditions and performance, plastic materials undergo many deformations and flow hence the study of mechanical, thermal and rheological behavior of polymers is of much significance for understanding and control of polymer formulation with compounding ingredients and subsequent processing and evaluation of their uses and applications.

Stress and Strain

Deformation of a material of a given dimension are generally caused by three types of forces (i) tensile; (ii) compressive and (iii) shear.

In general, stress is defined as the force applied, F, divided by area A, on which the force acts. For tensile and compressive stress, the area considered to define the stress is the cross-sectional area, A, perpendicular to the direction in which the force acts.

For calculation of shear stress (stress applied tangentially is called shear), however, the area considered is the surface area, A which is parallel to the direction of applied force F. In either case the stress concerned is equal to F/A.

The strain developed is expressed as the ratio of absolute deformation, ΔL is length divided by the original length of the unstressed sample L_0, i.e.,

$$\text{Strain} = \Delta L/L_0$$

A constant strain of any kind applied to an ideal elastic solid will deform it to a constant level of strain in equilibrium with applied stress and it will not strain or deform further with time.

5.6 MECHANICAL PROPERTIES

Strength, toughness, and ductility are all mechanical properties.

Stress-Strain Behavior

If a polymer sample is stressed under tension or compression, it shows strain. In a plot drawn between stress and % strain, the slope of the stress-strain curve evaluated at the origin is termed as the **elastic modulus** (in case of glassy polymers) and is taken to be a measure of the stiffness of the polymer being tested. For elastic polymers, the initial slope in stress-strain curve is called the **shear modulus**. The modulus may vary depending upon the type of stress. The point where stress-strain curves end is called the **fracture** point. The stress at fracture is called the **strength** of the polymer. The strain at fracture is called as the **elongation at break**, the larger the value of this, the more ductile is the polymer being tested. The area under the stress-strain curve is called the toughness of the polymer.

Strength

There is more than one kind of strength. There is **tensile** strength which is important for a material that is going to be stretched or under tension. A polymer sample has **compressional** strength if it is strong when one tries to compress it. A polymer sample has flexural strength if it is strong when one tries to compress or bend it. A sample's **torsional** strength is tested by twisting it. A sample has **impact** strength if it is strong when one hits it sharply and suddenly, like that as with a hammer.

Elongation

Elongation is a type of deformation. A polymer sample when deforms by stretching, becoming longer, the deformation is called elongation. It is usually measured in per cent, which is just the length the polymer sample is after it is stretched ($\angle$), divided by the original length of the sample ($\angle_0$), and then multiplied by 100:

$$\frac{L}{L_0} \times 100 = \%\ \text{elongation}$$

Two important elongations which are measured are ultimate elongation and elastic elongation.

Polymer Fracture

If a glassy polymer is stressed very rapidly or is stressed at a temperature that is much lower than its glass transition temperature, it tends to break or fracture in a brittle manner. An amorphous polymer tends to draw down in a homogenous manner if stressed at temperature above its glass transition temperature and shows large strains before fracturing. At intermediate temperatures and low rates of deformation, the polymer can yield somewhat before fracturing or fracturing in a ductile manner by neck formation.

At low strain (i.e., < 1%), the deformation of most polymers is elastic meaning that the deformation is homogeneous and full recovery can occur over a finite time. At higher strains, the deformation of glass polymers occur by either crazing, characteristic of brittle polymers or by a process called shear bonding which is the dominant mechanism for ductile polymers. Such deformations are not reversible unless the polymer is heated above its glass transition temperature.

Modulus

Modulus is measured to know how well a material resists deformation.

Toughness

Toughness is really a measure of the energy a sample can absorb before it breaks. Strength tells how much force is needed to break sample, and toughness tells how much energy is needed to break a sample. It is not necessary that just because a polymer sample is strong, it will be tough as well. From the following plots drawn between stress and strain, the difference between strength and toughness can be understood:

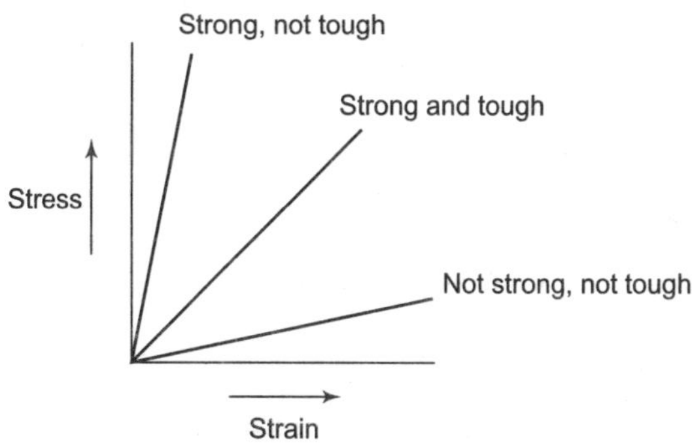

Fig. 5.8 Plots showing strength and toughness of polymeric materials.

Crazing

When some polymers such as polystyrene are deformed to a certain level, the critical strain (ε_0), what appears to be small cracks, develops in a direction perpendicular to the principal direction of deformation. These crazes reflect light and result in visual cloudiness or whitening of the sample.

A craze is a unique morphological feature of polymers and is morphologically different from a true crack. A craze, which can be from a nanometer to a few micrometers in thickness, consists of polymers microfibrils (0.6 to 30 nm in diameter) stretched in the direction of tensile deformation. The microfibrils are surrounded by void shape, which can represent as much as 90% of the total volume of the craze. This anisotropic morphology of a craze results in the scattering light.

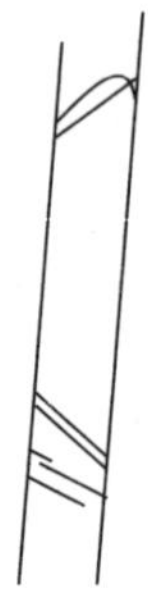

Fig. 5.9 Crazes Developed Perpendicular to the Tensile Direction

The time for the initiation of a craze depends upon many factors, including the magnitude of the applied stress, temperature and the presence of low molecular weight liquids which may act to promote craze development.

Although the reasons for craze initiation are still uncertain and several different theories have been proposed, it is recognized that crazes constitute the defects from which brittle cracks initiate.

Mechanical feature of a sample that has begun crazing is initiated by the breakdown of the fibrillar microstructure to form additional voids that grow slowly until some critical size has been reached. Beyond this point, the craze will rapidly propagate as a crack. As the crack propagates, crazes are formed at the crack tip and act to retard its advance.

Shear Banding

While some polymers such as polystyrene will readily craze when strained in tension, crazes may not develop in other polymers such as polycarbonate under identical conditions. Instead, these polymers will form regions of localized shear deformation. These regions are called shear bonds, which develop at angle of 45° to the stretch direction. Some polymers (e.g., SAN) can exhibit both modes of deformation. In general, shear bond formation is a dominant mode of deformation during tensile yielding of ductile polymers.

5.7 FLOW BEHAVIOR

5.7.1 Purely Viscous Flow

Rheology is the science of the deformation and flow of materials. Polymer systems exhibit interesting, unusual and difficult to describe (at least from the standpoint of traditional materials) deformation behavior. The simple and traditional linear engineering models, Newton's law (for flow) and Hooke's law (for elasticity) often just are not reasonable approximations.

The viscoelastic response is dramatically illustrated by "silly putty" (a silicone polymer) when it is bounced (stressed rapidly), it is highly elastic, recovering most of the potential energy it had before being dropped. If stuck on the wall, however, (stressed over a long time period) it will slowly flow down the wall, albeit with a high viscosity and will show little tendency towards recovering any deformation.

Purely viscous flow is defined as a deformation process in which all the applied mechanical energy is nonrecoverably dissipated as heat in the material; i.e., it is all converted to heat through viscous energy dissipation.

Purely viscous flow is in most cases a good approximation for dilute polymer solutions and often for concentrated solutions and melts where the stress on the material is not changing too rapidly, i.e., where it has a chance to approach on equilibrium flow situation.

The viscosity of a material expresses its resistance to flow under a mechanical stress. It is defined quantitatively in terms of two basic parameters, the shear stress τ and the shear rate or the rate of shear straining γ.

A coordinate system is established in which y axis is perpendicular to surfaces of constant fluid velocity; i.e., it is parallel to the velocity gradient. The z axis is mutually perpendicular to the others.

A flow field, in which the velocity and its gradient are everywhere perpendicular, is known as a viscometric flow. A fluid surface at 'y' moves with a velocity of 'u' is equal to dx/dt in the 'x' direction, while the surface at '$y+dy$' has velocity '$u+du$'. The displacement gradient, dx/dy, is known as the shear-strain.

$$\gamma = \frac{dx}{dy} = \text{shear strain (dimensionless)} \tag{5.13}$$

The time rate of change of shear strain r (the dot is Newton's rotation for the time derivative) is the so called shear rate. Since the order in which the mixed second derivative is taken is immaterial

$$\dot{r} = \frac{d}{dt}(r) = \frac{d}{dt}\left(\frac{dx}{dy}\right) = \frac{d}{dy}\left(\frac{dx}{dt}\right) = \frac{du}{dy}\left(\text{time}^{-1}\right) \tag{5.14}$$

Thus, an alternate definition of the shear rate is the velocity gradient d_u/d_y.

The shear stress is simply the force (in the direction of flow) per unit area normal to the y-axis:

$$T_{yx} = \frac{F\,(\text{in } x \text{ direction})}{A\,(\text{normal to } y \text{ direction})}\left(\frac{\text{force}}{\text{length}^2}\right)$$

The viscosity η is defined as the ratio of shear stress to shear rate

$$\eta \equiv \frac{T}{\dot{r}} \tag{5.15}$$

Relations between T (stress) and r (strain) Flow Curves

Newton's law of viscosity states that the shear stress is linearly proportional to the shear rate, the proportionality constant being the viscosity η.

$$T = \eta \dot{r}$$

(unit of η in cgs system is dyne/cm^2 or the poise).

Fluids that obey this hypothesis are termed **Newtonian fluid**. In flow, Newtonian fluids exhibit a direct proportionality between shear and shear rate. Newtonian behavior is generally found in gases and simple solutions of low molecular weight solutes such as water, sugar syrups, salt briness and most oils.

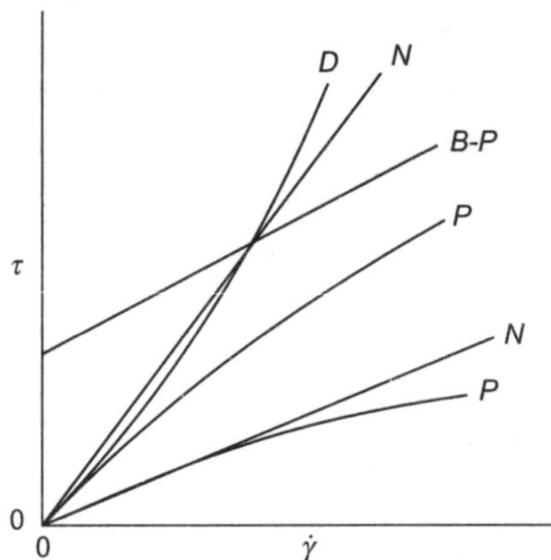

Fig. 5.10 Flow curves for different types of fluids-Newtonian (N); Psuedoplastic (P); Bingham Plastic (B-P)

It holds quite well for low molecular weight fluids. An arithmetic plot of T versus $\dot{r}$, a flow curve, is straight line through the origin with a slope η for a Newtonian fluid. Taking log of both sides of equation

$$\text{Log } T = \log \eta + 1 \log \dot{r} \tag{5.16}$$

Hence, log-log plot of τ versus $\dot{r}$, a logarithmic flow curve, will be a line of slope unity for a Newtonian fluid. This type of flow behavior would be expected for small molecules, where the structure and therefore the resistance to flow, does not change with the intensity of shearing.

Resistance to flow in **non-Newtonian fluids** is influenced by both temperature and shear rate or rate of flow. Non-Newtonian behavior can be divided into three broad categories:

(a) Time independent flow in which the resistnace to flow (apparent viscosity) is a function of the shear rate and/or shear stress,

(b) Time dependent flow in which flow properties are also influenced by the time of flow.

(c) Viscoelastic behavior which simultaneously displays both elastic and viscous properties.

Pseudoplastic Fluids

They exhibit shear rate thinning, in which apparent viscosity decreases with increasing shear rate.

Dilatant fluids

Dilatant flow behaviour is the opposite to that of pseudoplastics; i.e., apparent viscosity increases with increasing shear rate (shear thickening).

On logarithmic coordinates, dilatant flow curves have a slope greater than 1 and pseudoplastics a slope less than 1. Dilatant behavior is reported for certain slurries and implies on increased resistance to flow with intensified shearing. Polymer melts and solutions are invariably pseudoplastic; i.e., their resistance to flow decreases with the intensity of shearing.

The Power Law

The power law is very useful for understanding the flow behavior of many fluid systems that do not obey Newton's law.

$$T = K(\dot{r})^n \tag{5.17}$$

Where, 'n' is the flow behavior index, K is a constant (coefficient of viscosity). For Newtonian fluids $n = 1$. St. Vincent body fluid system after attaining a different degree of stress starts straining at any rate from zero to infinity and they disallow the stress to rise beyond the critical point.

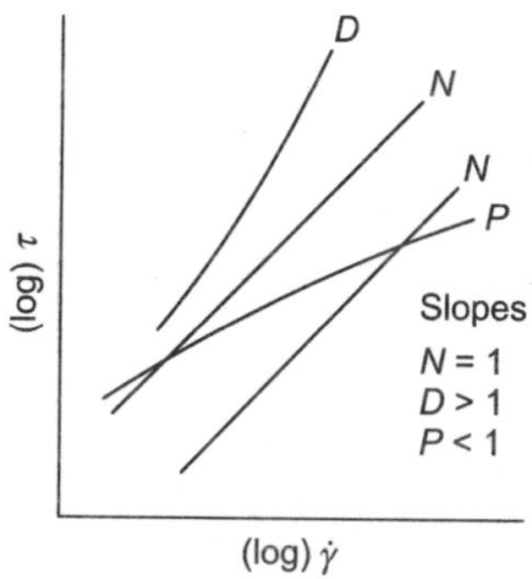

Fig. 5.11 Log flow curves for different types of fluids-Newtonian (N); Psuedoplastic (P); Bingham Plastic (B-P)

5.7.2 Time-dependent behavior

These types of non-Newtonian flow, though shear dependent, are time independent; i.e., as long as a constant shear rate or stress is maintained, the same viscosity will be observed at equilibrium.

Time Dependent non-Newtonian Fluids

Some fluids exhibit reversible time-dependent properties. Time dependent non-Newtonian fluids can be divided into two groups depending on whether the shear stress increases or decreases with time at a fixed shear rate and temperature. When sheared at a constant rate or stress, the viscosity of a thixotropic fluid will decrease over a period of time implying a progressive breakdown of structure. The former fluids are termed as antithixotropic or and the latter, thixotropic, provided that the time dependent effect is reversible. **Thixotropic** liquids exhibit a time-dependent response to shear strain rate over a longer period than that associated with changes in the shear strain rate. They may liquefy on being shaken and then solidify (or not) when this has stopped.

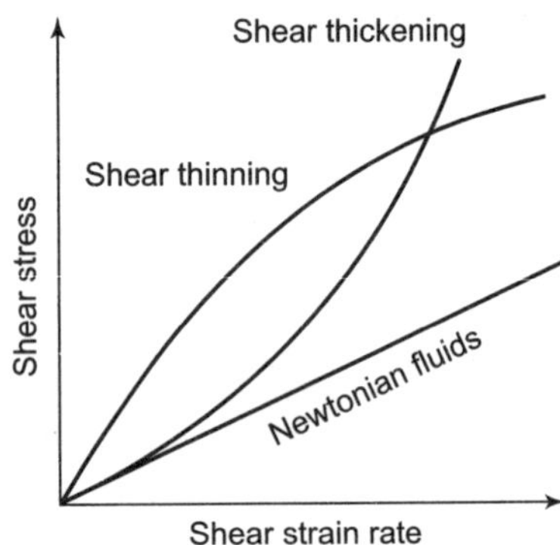

Fig. 5.12 Flow curves for different types of time dependent fluids-shear thinning (Thixotropic); shear thickening (Rheopectic)

Some fluids exhibit reversible time-dependent properties, however, when sheared at a constant rate or stress, the viscosity of a thixotropic fluid will decrease over a period of time implying a progressive breakdown of structure. The ketchup that splashes all over after a period of vigorous tapping is a classic example. Thixotropic behavior is important in the paint industry, where smooth, even application with brush or roller is required, but it is desirable for the paint on the surface to "set up" to avoid drips and runs after application.

The opposite sort of behavior is manifested by rheopectic fluids, for example, certain drilling muds used by the petroleum industry. When subjected to continuously increasing and then decreasing shearing time dependent fluids give the type of flow curves.

Once in a while, polymer systems will appear to be thixotropic or rheopectic but for chemically stable polymer melts or solutions, the equilibrium viscous properties are time independent.

5.7.3 Polymer Melts and Solutions

When the flow properties of polymer melts and solutions can be measured over a wide enough range of shearing, the logarithmic flow curves appear as follows:

It is generally observed that

(i) at low shear rates a lower Newtonian region is reached with a so called zero-shear viscosity η_0.

(ii) over several decades of intermediate shear rates, the material is pseudoplastics.

(iii) at very high shear rates, an "upper Newtonian" region with viscosity η is attained.

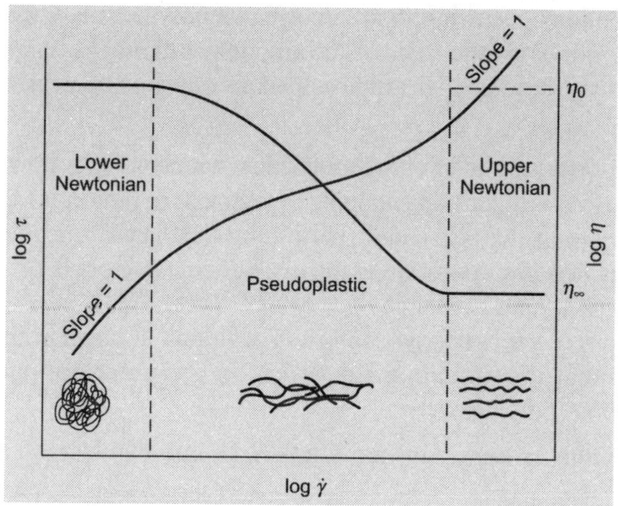

Fig. 5.13 Log flow curves for polymer melts and solutions showing stress-strain and strain-viscosity relationship

This behavior can be rationalized in terms of molecular structured. At low shear, the randomizing effect of thermal motion of chain segments overcomes any tendency towards molecular alignment in the shear field. The molecules are thus in their most random and entangled state, and have their greatest resistance to slippage or flow. As the shear is increased, the molecules will begin to entangle and align in the shear field, reducing their resistance to slippage past one another. Under severe shearing they will be pretty much completely untangled and aligned, and reach a state of minimum resistance to flow.

5.8 TRANSITIONS IN POLYMERS

Amorphous polymers can exhibit two distinctly different types of behavior. Some, like PMMA and PS are hard, rigid glassy polymers at room temperature while others like PB, PEA and polyisoprene are soft, flexible, rubbery materials. If however, PMMA and PS are heated up to 12°C, they exhibit typical rubbery properties and when a rubber ball is cooled in liquid nitrogen, it becomes rigid and glassy and shatters when an attempt is made to bounce it.

So, there is some temperature or narrow range of temperatures, below which a polymer is in a glassy state and above which it is rubbery. This temperature is known as glass transition temperature (T_g). T_g is a property of the polymer, and whether the polymer has glassy or rubbery properties depends on whether its application temperature is above or below its glass transition temperature.

5.8.1 Molecular Motions in Polymers

In order to understand the molecular basis for the glass transition, the various molecular motions occurring in an amorphous polymer mass may be broken into four categories on the basis of the amount of thermal energy required by them to occur: Motions 1–4 are arranged in order of decreasing activation energy:

1. translational motion of entire molecules, which permits flow.
2. cooperative wriggling and jumping of segments of molecules approximately 40–50 carbon atoms in length, permitting flexing and uncoiling.
3. motions of a few atoms along the main chain (five or six or so) or of side groups on the main chains.
4. vibrations of atoms about equilibrium positions, as occur in crystal lattices, except that the atomic centre is not in a regular arrangement in an amorphous polymer.

These motions or transitions are categorized into two classes:

Primary transitions: It is defined as one for which a discontinuity occurs in the first derivative of the Gibbs free energy.

According to first law of thermodynamics for a reversible, closed system, the Gibbs free energy can be expressed in differential form as a function of temperature and pressure $G(T, P)$, as

$$dG = -SdT + Vdp \tag{5.18}$$

S = entropy; V = system volume

The free energy may be differentiated with respect to temperature (at constant pressure) as

$$\left(\frac{\partial G}{\partial T}\right)_P = -S \tag{5.19}$$

and with respect to pressure at constant temperature as

$$\left(\frac{\partial G}{\partial p}\right)_T = V \tag{5.20}$$

In terms of describing transitions, in polymer, the most useful of the preceding relationship is the first derivative with respect to P which indicates that a first order transition should occur as a discontinuity in volume.

At T_m, first order transitions occur, i.e., motions 1 and 2.

Second Order Transitions: Second order transitions mean that a discontinuity should be observed in the second derivatives of Gibbs free energy

$$\left(\frac{\partial^2 G}{\partial T^2}\right)_P = \left(\frac{\partial G}{\partial T}\right)_P \tag{5.21}$$

$$\left(\frac{\partial^2 G}{\partial p^2}\right)_T = \left(\frac{\partial V}{\partial P}\right)_T \tag{5.22}$$

and

$$\left[\frac{\partial}{\partial T}\left(\frac{\partial G}{\partial p}\right)_T\right]_P = \left(\frac{\partial V}{\partial T}\right)_P \tag{5.23}$$

The T_g is thought to be that temperature at which there is only sufficient energy available for motion 3 and 4 (secondary transitions) which means that a discontinuity should be observed in the second derivatives of the Gibbs free energy. Glass transition or second order transition is not a phase transition and almost each polymer is characterized by a specific T_g appearing well below its T_m.

Glass state (Brittle Plastics)	Rubbery or Viscoelastic state (Tough plastics and rubbers)	Viscoelastic state (Polymer melts)
T_g (Glass Transition Temp.)		T_f (Flow Temp.)

5.8.2 Determination of T_g

A common method is to observe the variation of some thermodynamic property with T, for example, the specific volume.

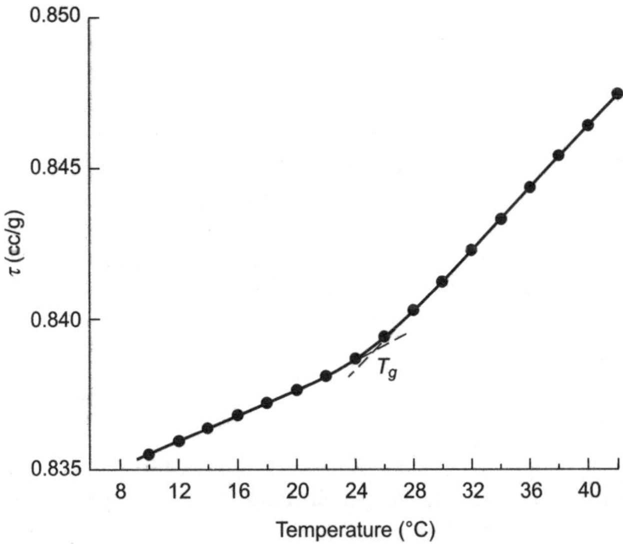

Fig. 5.14 Plot between specific volume and temperature showing change in slope value at glass transition temperature

The slope of the v vs. T plot increases above the T_g. The value of T_g determined in this fashion will vary slightly with the rate of cooling or heating. This reflects the fact that long entangled polymer chains cannot respond instantaneously to charges in temperature and illustrates the difficulty in making thermodynamic measurements on polymers. There is never observed a sharp break in the property but this is of little practical importance because T_g can always be established within a couple of degrees by extrapolation of the linear regions. Other properties such as refractive index, may also be used to establish T_g.

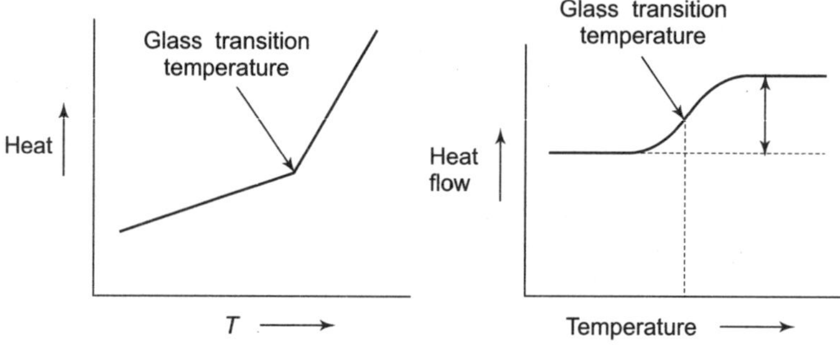

Fig. 5.15 Plots between heat flow and temperature showing change in slope value at glass transition temperature

Factors Influencing T_g

In general, the glass transition temperature depends on five factors:

1. The free volume of the polymer V_f: Free volume is the volume of the polymer mass not actually occupied by the molecules themselves that is $V_f = V - V_s$, where V is the specific volume of the polymer and V_s the volume of the solidly packed molecules. The higher the V_f, the more room the molecules will have to move around and lower the T_g. It has been established that for all polymers

$$V_f/V = 0.025 \text{ at } T_g$$

2. The attractive forces between the molecules: The more strongly the polymer moles are bound together, the greater the thermal energy required to produce motion. Since the solubility parameter δ is a measure of intermolecular forces, T_g increases with δ.
3. The internal mobility of the chains, i.e., their freedom to rotate about bonds: Potential energy is a function of rotation angle about a bond in a polymer chain. The minimum energy configuration (arbitrarily chosen as $\theta = 0$) is the position where the larger substituents (the rest of the chain) are as far away from each other as possible.

   In a graph between potential energy and rotation about a bond in a polymer chain backbone, the minimum energy configuration is the position where the largest substituents (the rest of the chain) are as far away from each other as possible. T_g increases with E_0, higher the E_0, higher will be the T_g.

4. The stiffness of the chains: Chains that have difficulty in coiling and folding will have higher T_g's. The stiffness usually goes hand in hand with high E_0, so it is difficult to separate the effects of 3 and 4.

   Chains with parallel bonds in the backbone (ladder type polymers), for example, polyimides and those with highly aromatic backbones have extremely stiff chains showhigher T_g's. This makes them mechanically useful at elevated temperatures, but also very difficult to process.

5. The chain lengths: As do many mechanical properties of polymers, the T_g varies according to the empirical relation

$$T_g = T_g^{\infty} - \frac{C}{x}$$

where C = constant for particular polymer
T_g^{∞} = asymptotic value of the T_g at infinite chain length
x = chain length

In most cases, for $x > 500$ (the range of commercial interest for most polymers),

$$T_g \cong T_g^{\infty}$$

5.9 POLYMER DEGRADATION

Now the use of synthetic polymers has accelerated to the extent that the disposal of the polymeric materials has become very difficult. Polymer degradation means a reduction in the molecular weight of the polymer.

Types of Degradation

(i) Chain end degradation
(ii) Random degradation

6

POLYMER CHARACTERIZATION

Learning Objectives

After studying this chapter, you will know:

- How polymers are identified and characterized?
- About common spectroscopic techniques used for polymer characterization.
- How polymer solution viscosity is measured?
- About the techniques used for molecular weight determination.
- How is morphology of polymer molecules studied?
- How is thermal analysis of a polymer done?
- How are mechanical properties of a polymer sample tested?

Chapter Contents

• Identification and Characterization of Polymers • Physical and chemical analysis of polymers • Spectrometric techniques used for structure determination • Various chromatographic techniques used for molecular weight determination • Surface structure or morphology • Thermal analysis • Evaluation of mechanical properties

6.1 IDENTIFICATION AND CHARACTERIZATION OF POLYMERS

Identification of an unknown sample of a plastic, rubber or a fiber may be required for a variety of reasons. Many of these polymers carry their trade name only and their detailed structure is not disclosed. In such cases, it may be necessary to analyze the polymer sample and identify it. A variety of lab techniques are used to determine the properties of polymers. Techniques such as wide angle X-ray scattering, small angle X-ray scattering, and small angle neutron scattering are used to determine the crystalline structure of polymers. Gel permeation chromatography (GPC) is used to determine the number average molecular weight, weight average molecular weight, and polydispersity. FTIR is used to determine composition. Thermal properties such as the glass transition temperature and melting point can be determined by differential scanning calorimetry (DSC) and dynamic mechanical analysis (DTA). Pyrolysis followed by analysis of the fragments is one more technique for determining the possible structure of the polymer.

Preliminary Tests

The preliminary tests include visual examination, cutting, effect of heat and burning, solubility and fusion tests.

Visual Examination

A visual examination of the sample may reveal useful information. For example, the presence of a hard, inflexible flash line would indicate a thermoset molded material. Similarly, the presence of a gate scar would indicate an injection molded material. However, this should not be confused with the gate scar on a thermoset transfer molded material. The physical form of the sample, i.e., whether granules, film, sheet or fiber and its flexibility or rigidity would give some indication of its identity. The color of the product can also be used for identification of a thermoset material. For example, a pastel shade would rule out a phenolic and probably indicate that it is a urea or melamine-formaldehyde plastic material.

Cutting

The cutting of a plastic material with a pen knife can also provide some information. For example, one can differentiate between crystal clear cellulose acetate and poly(styrene). Thermoset and thermoplastic material can also be differentiated in this way. The ivory like cut of casein and cast phenolics is noteworthy.

Heating Tests

A small amount of the material to be tested is taken in a deep spatula and heated on a small bunsen burner. The color of flame, ease of burning, whether the burning continues after removal from the flame and so on, all give an indication of the possible identity of the material. Precautions should be taken during heat test as some materials, like cellulose nitrate, explode or burn away rapidly.

A second heating test can be conducted with the help of a clean copper wire. The wire is first heated in a clear Bunsen flame and then touched with a small quantity of the material. It is heated again and the color of the flame is noted. Blue and green color indicates the presence of halogens in the composition, that is, chlorine, fluorine and rarely, bromine. Presence of poly(vinyl chloride) or its co-polymers, poly(vinylidine chloride), poly(tetrafluorethylene), chlorinated rubber, rubber hydrochloride or cellulose acetate containing a plasticizer like tricresyl phosphate is indicated by this test.

A third heating test is carried out by heating a small sample in a hard glass tube. The gas evolved is condensed in another tube and very carefully smelled. This can then be compared with the gas evolved from a known polymer.

Solubility

For thermoplastic polymers, a scheme of identification based on their solubility behavior has been put forward. The chart shows six different examples where at least two polymers are found by the same solubility route. Other similar examples will certainly be found if this chart is further generalized. In case of ambiguity, the solubility test has to be complemented by calorimetric and/or spectroscopic analyses.

6.2 CHEMICAL ANALYSIS OF POLYMERS

Fusion Test

The metallic sodium fusion test, can show the presence or absence of nitrogen and halogens. Similarly, the potassium nitrate/potassium carbonate fusion test will indicate the presence or absence of phosphorous in the material under examination. Thus, one can easily distinguish between Nylon and Polyesters.

From the preliminary tests, a great deal of information can be gathered regarding the possible identity of the unknown polymer. If the polymer does not show rubber like elasticity, an elastomer is ruled out. Further, if on heating it does not melt or flow, a thermosetting polymer is indicated. If it does melt, a thermoplastic polymer is indicated.

Carbon, Hydrogen and Nitrogen Analysis

If the fusion test indicates the presence of nitrogen, a quantitative analysis of the compound for C, H and N is carried out and if it tallies with that of a known polymer the identity of the polymer material is known with some amount of certainty. If other elements besides C, H and N are present, such as P or Cl, they have also to be taken account of in the calculation.

6.3 SPECTROSCOPIC METHODS

6.3.1 Infrared Spectroscopy

IR spectroscopy is a very simple analytical technique. First, the material to be analyzed is put in some form that can be put into the infrared spectrometer. This is usually accomplished by casting a film on a sodium chloride (table salt) salt plate, or by grinding the material up with potassium bromide, KBr, and making a pellet out of it. These salts are used because they are invisible to IR. There are other ways to make a sample, but these are the most common when dealing with polymers. Next, the sample is placed into the sample chamber of spectrometer and in less than a minute an IR spectrum of polymer is taken.

Infrared, or IR, spectroscopy is one type of vibrational spectroscopic technique where molecular vibrations are analyzed. To fully understand IR spectroscopy, one must first understand the principles of simple harmonic motion.

Imagine two spheres, or masses, connected with a spring. This is what is known as a *simple harmonic oscillator*. Once set into motion, the sphere will oscillate, or vibrate back and forth on the spring, at a certain frequency depending on the masses of the spheres and the stiffness of the spring.

A sphere with a small mass is lighter and easier to move around than one with a large mass. Therefore, smaller masses oscillate at a higher frequency than larger masses. A very stiff spring, like a bedspring, is hard to deform and quickly returns to its original shape when the deforming force is removed. On the other hand, a weak spring is easily deformed and takes much longer to return to its shape. Therefore, a stiffer spring will oscillate at a higher frequency than a weak one. A chemical bond between two atoms can be thought of as a simple harmonic oscillator. The bond is the spring, and the two atoms, or groups of atoms, connected by the bond are the masses. Every atom has a different mass, and single, double and triple all have different level of stiffness, and therefore each combination of atoms and bonds has its own characteristic harmonic frequency. When an object is vibrating at a certain frequency and encounters another vibration of exactly the same frequency, the oscillator will absorb that energy. Take a guitar string, for example. If the G-string is plucked and set on as "a-vibration", it would make that beautiful "G" sound. If the D-string is then plucked while holding it at the fifth fret, it would also make the "G" sound. But if one looks at the strings closely, one would see that not only would the D-string be vibrating, but the G-string would also be vibrating because some of energy from the vibrating D-string was transferred to the G-string. This is also true of vibrating molecules, except plucking a G-string won't affect the vibrations of chemical bonds.

If at any temperature above absolute zero, all the "eensy-weensy" little simple harmonic oscillators that make up any molecule are vibrating vigorously, infrared light just happens to be in the same frequency range as a vibrating molecule. If some IR light hits a vibrating molecule, it will absorb those frequencies in the light which exactly match the frequencies of the different harmonic oscillators that make up that molecule. When this light is absorbed, the little oscillators in the molecule will continue to vibrate at the same frequency, but since they have absorbed the energy of the light, they will have a larger amplitude of vibration. This means that the "springs" will stretch further than before the light was absorbed. The remaining light which was not absorbed by any of the oscillators in the molecule is transmitted through the sample to a detector, and a computer will analyze the transmitted light and determine what frequencies were absorbed. Each group of atoms in a molecule vibrate at its own frequency, and there are six different ways the molecules can vibrate. Each of these vibrations can occur in any molecule, but they all occur at different frequencies.

Earlier it was only possible to get good data by hitting the molecule with only one frequency of IR light at a time. This took very long because there are a lot of frequencies and to get a good spectrum, many scans must be taken. But now with the amazing Fourier Transform Algorithm, one can hit the molecule with every frequency of IR light at once, and get a perfect spectrum in only a fraction of the time.

$$F(v)=\int_{-\infty}^{\infty} f(t)e^{-i(2\pi)vt}\,dt \quad \text{and} \quad f(t)=\int_{-\infty}^{\infty} 2\pi F(v)e^{i(2\pi)vt}\,dt$$

Fourier Transform Algorithm The characterization of a polymer requires several parameters which need to be specified. This is because a polymer actually consists of a statistical distribution of chains of varying lengths, and each chain consists of monomer residues which affect its properties. In a complex polymer molecule, the number of infrared transitions might be expected to be too great to deal with, but fortunately this does not happen as a great many of these are degenerate, i.e., of the same energy. This allows the recognition of vibrational bands representing particular types of bonds and functional groups. They appear at similar positions regardless of the specific compound in which they occur. At the high frequency end of the infrared spectrum, the bands represent the individual bonds in the molecule while at the low frequency end of the spectrum, the vibrational bands are more characteristics of the molecule as a whole. This region is commonly known as the 'finger print region'. Advantage is taken of the occurrence of this complexity in the identification of polymers. The spectrum of the unknown sample is matched against that of an authentic sample, peak by peak, and the identity of the compound is established. It is unlikely that two compounds, except the optical enantiomorphs, will give the same infrared spectrum. Table 6.1 shows some of the most typical infrared absorptions and where they show up in a spectrum.

Table 6.1 Typical Infrared Absorptions

Vibration	*Frequency range (cm–1)*
carbonyl (C=O) stretch	1870 – 1650
Alcohols	
O-H stretch	3640 – 3250
C-OH stretch	1160 – 1030
C-OH in-plane bend	1440 – 1260
C-OH wag	700 – 600
Alkanes	
C-H stretch	2980 – 2850
CH_2 wag	1470 – 1450
CH_2 rock	740 – 720
CH_3 wag	1390 – 1370
CH_3 twist	1470 – 1440
Alkenes	
$=CH_2$ stretch	3040 – 3010
$=CH_2$ wag	950 – 900
C=C stretch (*cis* isomer)	1665 – 1635
C=C stretch (*trans* isomer)	1675 – 1665
Amines	
N-H stretch	3460 – 3280
NH_2 wag	1650 – 1590
C-N stretch	1190 – 1130
C-N-C scissor	510 – 480
Esters	
C-O-C asymmetrical stretch	1290 – 1180
O-C-O scissors	645 – 575
Nitro compounds	
NO_2 symmetrical stretch	1570 – 1550
NO_2 asymmetrical stretch	1380 – 1360
NO_2 scissors	650 – 600
NO_2 rock	530 – 470
Sulfones	
SO_2 symmetrical stretch	1170 – 1120
SO_2 asymmetrical stretch	1360 – 1290
SO_2 scissors	610 – 545

6.3.2 NMR Spectroscopy

NMR is also a characterization technique where a sample is immersed in a magnetic field and hit with radio waves. Pulsed Fourier transform NMR in which a high power microsecond-pulse of radio frequency energy sets all the carbon nuclei into resonance is used to confirm the chemical structure of polymeric materials. The data are recorded as the subsequent decay of the resonances with time, which is the Fourier transform of the desired spectrum. To decode the resonance (signals), a decoder called the Fourier Transform algorithm is required. It is a complex equation that translates the language of the nuclei into something we can understand. Here's the algorithm, in case you are curious.

$$F(v) = \int_{-\infty}^{\infty} f(t) e^{-i(2\pi)vt} \, dt \quad \text{and}$$

$$f(t) = \int_{-\infty}^{\infty} 2\pi F(v) e^{i(2\pi)vt} \, dt$$

These two equations translate the NMR signal from one form to another. Signals are analyzed to determine many different things about the molecule and its surroundings, such as the molecule's structure.

As the positively charged nucleus spins, this moving charge creates a magnetic moment and it can be imagined as a spinning subatomic bar magnet. When no magnetic field is present, these tiny magnets are aligned randomly, but when they are placed in a homogenous magnetic field, the magnetic moments will line up with the magnetic field. Even though the magnetic moments are aligned by the magnetic field, the nuclear spin is not simple and flat like the spin of a merry-go-round. The thermal motion of the molecule creates a torque which makes the magnetic moment "wobble" like a child's top. When the radio waves hit the spinning nuclei, they tilt even more, and even flip over sometimes. When the magnetic moment is tilted away from the applied magnetic field, some of the magnetic moment is detectable perpendicular (90°) to the applied magnetic field.

Different nuclei resonate at different frequencies. This means that a carbon atom must be hit with a different frequency radiowave than a hydrogen atom to get it to flip. It also means that similar atoms in different environments, such as a hydrogen attached to an oxygen and a hydrogen attached to a carbon, flip at different frequencies. By seeing at which frequencies these different nuclei flip, one can determine how a molecule is put together, as well as many other interesting properties of the molecule. Two different atoms, like carbon and hydrogen, resonate at different frequencies because they are different, but two similar atoms in different environments resonate at different frequencies as well due to shielding. The electrons which surround the spinning nuclei are also charged and spinning that create a magnetic field which is in opposition to the applied magnetic field. This decreases the magnitude of the applied magnetic field which reaches the nuclei. In other words, the electrons "shield" the nuclei from the full magnetic field. Because the resonant frequency of a nuclei depends on the strength of the magnetic field it experiences. In the NMR spectrometer there is a coil of wire which surrounds the sample. A magnet moving in a wire coil causes a charge to flow in the wire. So when the magnetic field of the nucleus flips, it causes a current in the wire which the computer attached to it can detect.

There are many different nuclei which can be observed by NMR spectroscopy: ^{1}H (proton), ^{13}C (carbon 13), ^{15}N (nitrogen 15), ^{19}F (fluorine 19), and many others. ^{1}H, and ^{13}C are used the most often. ^{1}H is described here because it represents of NMR spectroscopy. Table 6.2 contains just a few of the many different chemical shifts (given in parts per million, or ppm) possible in ^1H NMR and the protons associated with those shifts.

Table 6.2 Different Chemical Shifts in ^1H NMR Spectrum

Proton	***Chemical shift (ppm)***		
ethane (H_3C-CH_3)	1.25		
propane (H_3C-CH_2-CH_3)	1.17		
cyclopropane (C_3H_6)	0.22		
cyclopentane (C_5H_{10})	1.40		
cyclohexane (C_6H_{12})	1.35		
toluene (H_3C-C_6H_5)	2.87		
styrene (H_2C=CH-C_6H_5)	5.09		
styrene (H_2C=CH-C_6H_5)	6.68		
Aromatics	*ortho*	*meta*	*para*
benzene (C_6H_6)	7.54	7.54	7.54
toluene (H_3C-C_6H_5)	7.16	7.16	7.16
styrene (H_2C=CH-C_6H_5)	7.24	7.24	7.24
bromobenzene (Br-C_6H_5)	7.49	7.14	7.24
phenol (HO-C_6H_5)	6.77	7.13	6.87

6.4 MOLECULAR WEIGHT DETERMINATION

Physical Methods—These methods depend on physical properties.

freezing point depression—gives $M(n)$, the number average molecular weight.

boiling point elevation—gives $M(n)$, the number average molecular weight.

osmotic pressure—gives $M(n)$, the number average molecular weight.

For freezing point depression, boiling point elevation, and osmotic pressure methods to work, the polymer must be soluble, and the test samples must be made from dilute solutions.

Light Scattering—Light scattering gives $M(w)$, the weight average molecular weight. Light scattering occurs because the dimensions of the polymer molecules are on a similar size scale as the wavelength of the light.

Sedimentation gives the $M(z)$, z average molecular weight, and is used primarily in biochemistry where polydispersity values are ~1. Sedimentation is not an option if the polymer samples are polydisperse.

Viscosity gives the Mv, the viscosity average molecular weight. Mv is a quick method, but there are two constants in the calculation, K and a, which must be known. To find out what these constants are, it is necessary to take known standards of the polymer you wish to test, and run viscometry experiments.

Relative Viscometer

Solution viscosities are determined by the viscosity of the sample relative to the reference solvent. The relative viscometer measures the solvent and sample viscosity simultaneously, therefore errors due to temperature fluctuation and solvent variations are avoided.

The Relative Viscometer is ideally suited for measuring dilute polymer viscosities. It provides faster analysis and greater precision than is obtainable with the conventional glass tube viscometers (Ubbelohde or Cannon-Fenske) it replaces. The principle of operation is based on measurement of pressure drops due to the continuous forced flow of solvent and sample through two stainless steel capillary tubes placed in series. The pressure drop across each capillary tube obeys Poiseuille's Law.

$$P = \eta Q R$$

P is the pressure drop as measured by a differential pressure transducer (DPT). Q is the flow rate, R is the tube resistance, and η is the viscosity. The sample solution is loaded into the sample loop via the syringe pump and then pushed into capillary 2 when the injection valve is turned. A steady state condition is reached when the sample solution completely fills capillary 2, solvent remaining in capillary 1 at all times. The relative viscosity of the sample solution is determined simply and directly by the ratio of the pressure drops.

$$\eta_r = \frac{\eta}{\eta_0} = \frac{R_1 P_2}{R_2 P_1} = K \frac{P_2}{P_1}$$

From the measured relative viscosity, all other solution viscosity measurements can be calculated.

6.4.1 Measuring Polymer Solution Viscosity

Polymer solution is taken in a viscometer that looks like the one in the picture, and time taken for a given volume of the solution to flow through it between the lines a and b as shown in the picture below, is measured:

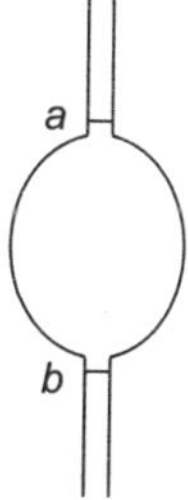

Fig. 6.1 A Viscometer

The time taken for the solution level to drop is called an efflux time. Efflux time for solutions of polymer at varying concentrations and that of the pure solvent, with no polymer dissolved in it are measured. Let the efflux time of pure solvent and polymer solution be t_0 and t. The relative viscosity is calculated by dividing t by t_0:

$$\frac{\text{efflux time of solution}}{\text{efflux time of the solvent}} = \text{relative viscosity}$$

$$\frac{t}{t_0} = \eta_r$$

Specific viscosity is calculated as:

$$\frac{\text{(efflux time of solution – efflux time of solvent)}}{\text{efflux time of the solvent}} = \text{specific viscosity}$$

$$\frac{t - t_0}{t_0} = \eta_{sp}$$

The reduced viscosity is calculated by dividing the specific viscosity by the concentration of the polymer solution in question:

$$\frac{\text{specific viscosity}}{\text{concentration}} = \text{reduced viscosity}$$

$$\frac{\eta_{sp}}{c} = \eta_{red}$$

The plot drawn between reduced viscosity at each concentration at y-axis, and concentration on the *x*-axis, gives a straight line. The slope of this line gives the value of constant '*k*' and if this line is extrapolated to zero concentration, the intercept at *y*-axis gives the value of intrinsic viscosity. (The intrinsic viscosity is a hypothetical construct.)

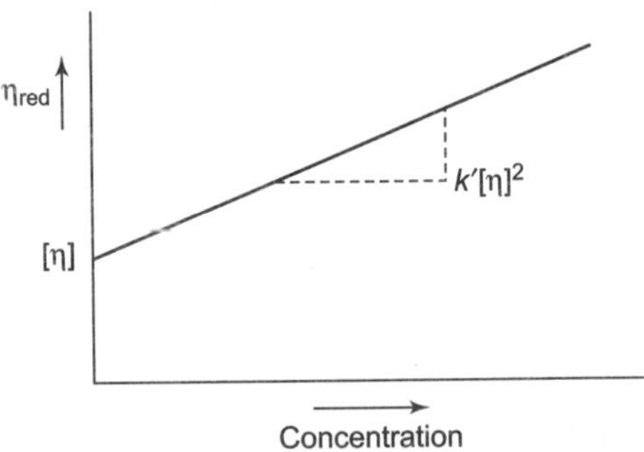

Fig. 6.2 A Plot of Reduced Viscosity vs. Concentration

The intrinsic viscosity is an important number, because it is the one that will tell us the molecular weight, momentarily. A second way to calculate intrinsic viscosity is to calculate inherent viscosity first,

$$\frac{\text{ln (relative viscosity)}}{\text{concentration}} = \text{inherent viscosity}$$

$$\frac{\ln \eta_r}{c} = \eta_{inh}$$

and then drawing a plot between inherent viscosity and concentration as shown below:

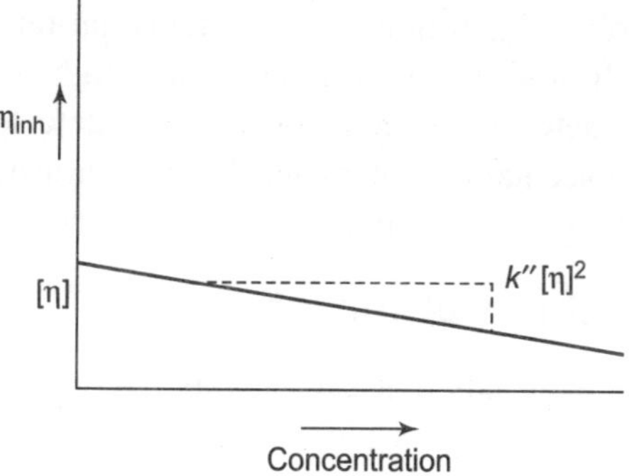

Fig. 6.3 A Plot of Inherent Viscosity vs. Concentration

Intrinsic viscosity is calculated using both methods. The plots from both methods are put together to get a plot that looks like that shown below, with the two lines meeting at their common intercept:

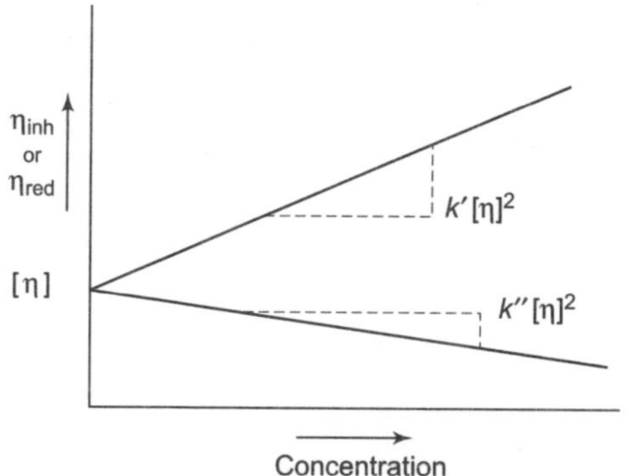

Fig. 6.4 A Plot of Inherent/Reduced Viscosity vs. Concentration

The value of k′ - k″ should equal 0.5. The molecular weight of a polymer sample can be calculated by a simple equation:

$$[\eta] = K'M^{a}$$

This is called the Mark–Houwink equation. *M* is the viscosity average molecular weight and *K′* and *a* are the Mark–Houwink constants. There is a specific set of Mark–Houwink constants for every polymer–solvent combination. The values of these constants are required for any polymer–solvent combination in order to get an accurate measure of molecular weight. Only dilute solutions are used to do this kind of experiment as the polymer molecules might get close enough together to interact with each other if the solutions are too concentrated. This causes the viscosity to increase in ways that the above equation does not apply very well, as a result, accurate measurements cannot be made. Thus, the technique is called dilute solution viscometry.

There are several techniques by which the molecular weight of a polymer sample can be determined. Gel Permeation Chromatography (GPC) / Size Exclusion Chromatography (SEC) are used for the characterization of natural and synthetic polymers, copolymers, proteins, protein conjugates, aggregates and excipients. In a single experiment, static light scattering, dynamic light scattering, viscometer, triple detector and tetra detector technologies allow to obtain absolute molecular weight, molecular size, intrinsic viscosity, conformation, structure, branching, aggregation and copolymer composition.

6.4.2 Gas Chromatography

GC is useful for materials that can be vaporized. The three main parts of the GC are in the injector, column and detector. The sample is injected by a syringe into the injector, and it immediately vaporizes. For capillary GC systems a splitter permits only a small portion of the sample to enter the column. The carrier gas, usually helium, pushes the sample onto the column. The columns are packed with a liquid coating over an inert support such as silicone oil or firebrick. The liquid packing must be high boiling and have a small vapor pressure of less than 1 mm at the maximum temperature used so that it is not removed from the column.

For a capillary column system, the coating materials are coated directly on the walls of the column. A capillary column is a more efficient system because a longer column will fit into a given volume of GC interior, owing to the smaller diameter. Capillary column lengths of 100 to 150 feet are common.

The materials are separated on the column by selective interaction of one material more than the other. The material that interacts the most, moves the slowest. The interactions can be absorption or any type of chemical interaction. Even if two substances have the same boiling point, they can usually be separated, due to the principle that deals with how well the column absorption is.

The two types of detectors for the GC are thermoconductivity (TC) and flame ionization (FID.) The T.C. detector measures the difference in the thermal conductivity of the sample vs. a reference sample using a Wheatstone bridge to measure the imbalance. The FID system uses a hydrogen flame to burn the sample and it measures the amount of materials by this method. FID systems are usually very much more sensitive than TC by a factor of 1000 or more. Other detectors include electron capture, Halls, or other specific detectors. The ability to vary the temperature of the column permits a greater range of separation.

6.4.3 Liquid Chromatography

LC or High Performance Liquid Chromatography (HPLC) is useful for materials that dissolve in solvents. Just as with GC, there is the injector, the column, and the detector. Rather than using a gas, one uses a solvent as the carrier media.

A liquid pump is needed. The pump should be capable of providing uniform, pulse free, flow of 0.1 to 10 ml/min at pressures up to 6,000 psi. A loop injector is used to introduce the sample into the system.

Reverse phase columns are prepared by chemically bonding the organic phase to the inert support. This prevents the solvent from washing the organic phase form the packing.

The three main types of detectors include refractive index (RI), ultraviolet (UV), and electrochemical. The RI detector measures the difference between a reference solvent and a sample peak as it elutes from the column. Differences in the 7th decimal place are important, and because of this, temperature control and solvent uniformality are important. The UV detector covers the range of 200 to 700 nm. It is more sensitive than RI. It has two other advantages:

(1) It can be tuned to the exact frequency of the material.
(2) It is insensitive to changes in solvent composition. The electrochemical detector works for charged species or ions.

Typical solvent pairs for liquid chromatography include: water/methanol; water/acetonitrile; chloroform/hexane; and others may be used.

6.4.4 Gel Permeation Chromatography/ (Size Exclusion Chromatography)

GPC is used to determine the molecular weight of polymers. It can use the same system as used by LC. The pump is the same. Either UV or RI detector can be used, but usually an RI detector is used. The solvent of choice is THF, but toluene or *m*-cresol are alternatives. Almost any solvent can be used that will dissolve polymers. The RI of the solvent must be different from the RI of the polymer to be analyzed. Solutions with 0.25 to 0.5 per cent (w/v) are used for the analysis.

Crosslinked polystyrene can be used for the packing material. Packing materials must be insoluble in the solvent used, so a solvent for the experiment must be chosen which does not dissolve the packing material. This can be a tiresome issue, as generally, polymers are insoluble in many solvents. The two main types of detectors for GPC include TC and FID. The TC detector measures the difference in thermoconductivity of the sample vs. a reference sample using a Wheatstone bridge to determine the imbalance (the difference). The flame ionization detector uses a hydrogen flame to burn the sample and it measures the amount (mass) of materials by this method. FID systems are usually much more sensitive than TC systems by a factor of 1000 or more.

After running a GPC experiment a graph of detector response as a function of time is made.

One can run standards of known molecular weight and determine where they appear on a graph of signal vs. time. From these, one can determine a function of molecular weight vs. time, and from this one can convert a signal vs. time graph to a signal vs. molecular weight. From the signal vs. molecular weight graph, one may calculate molecular weight (number average, weight average, etc.).

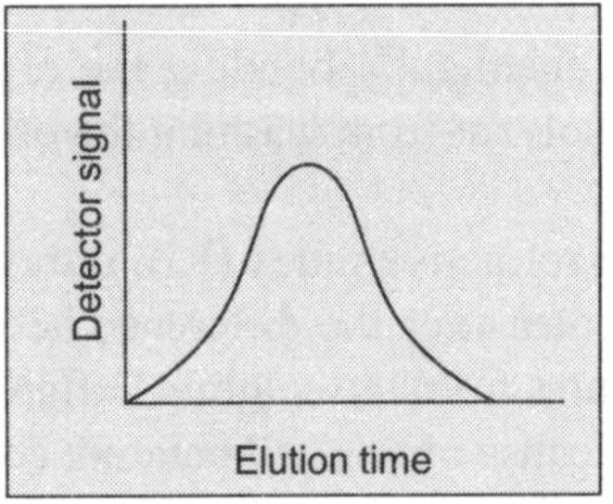

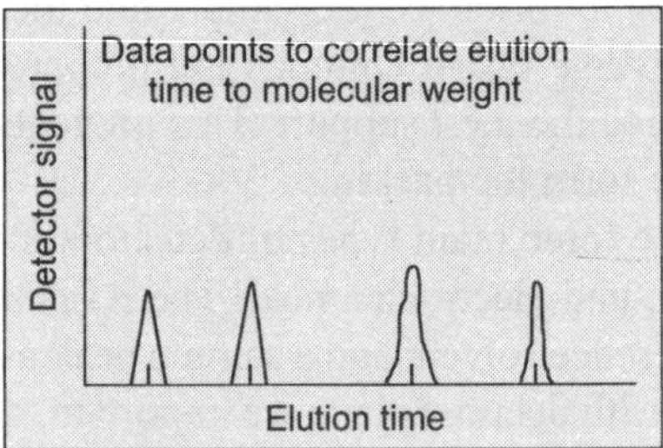

Fig. 6.5 GPC Curves

6.4.5 Low Angle Light Scattering (LALS)

LALS is the only GPC detection method that provides absolute molecular weights without extrapolation or correction. Unlike multi-angle light scattering techniques that can only determine molecular weight by extrapolation, LALS directly measures molecular weight for a large range of synthetic polymers, natural polymers and proteins.

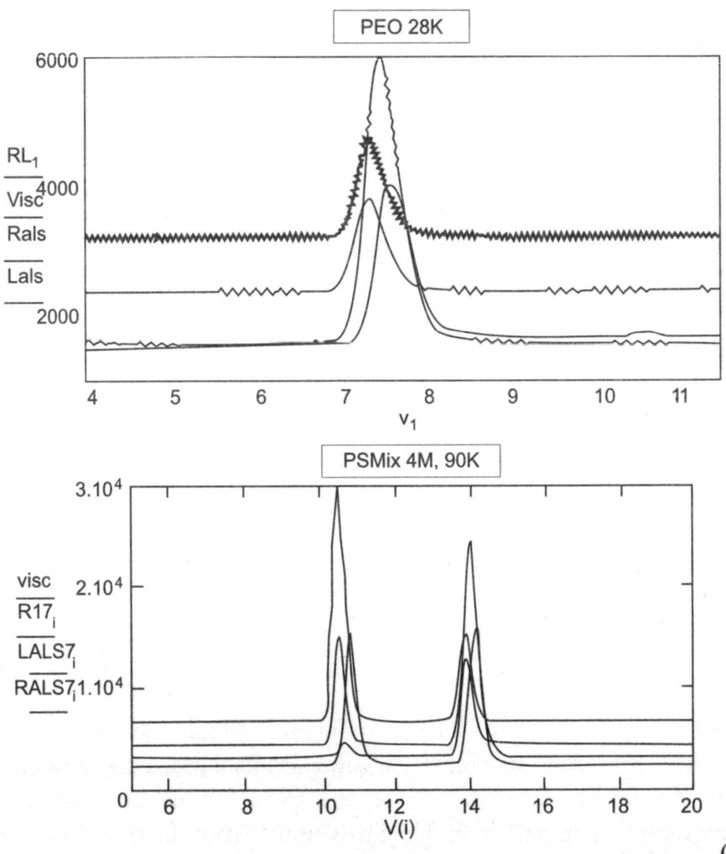

(Contd.)

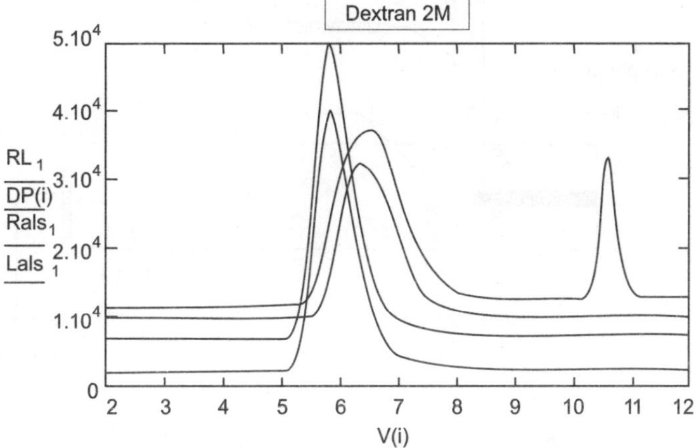

Fig. 6.6 LALS Chromatograms

6.4.6 Dynamic Light Scattering

DLS is the fast and accurate way to characterize your biomolecules in solution. One require only a small volume of sample (10 μl) which is recoverable. The sample is introduced into the sample cell which is placed in the temperature controlled cell holder. The software controlled instrument then makes the measurement and the results are displayed in seconds. A complete analysis from start to finish can be accomplished in minutes.

6.4.7 Flow Injection Polymer Analysis

Flow Injection Polymer Analysis (FIPA) is a new method for high through-put screening and routine process and quality control. FIPA is a fast, precise, accurate and reliable technique for measuring molecular weight, molecular size, intrinsic viscosity and per cent polymer without the need for elaborate sample preparation.

The control of the polymerization process requires fast turn around time for sequential measurements of the extent of reaction. FIPA measures polymer concentration, molecular weight and intrinsic viscosity of a solution or emulsion reactor sample in less than five minutes. The typical system consists of a pump, an injector, a FIPA column and a Triple Detector Array (TDA). The TDA combines a light scattering detector, viscometer detector and refractometer in a single integrated unit with fixed inter-detector volumes and temperature control.

6.4.8 MALDI

To know exactly what the molecular weight distribution is in a polymer sample, matrix-assisted laser desorption/ionization mass spectroscopy (MALDI) is used.

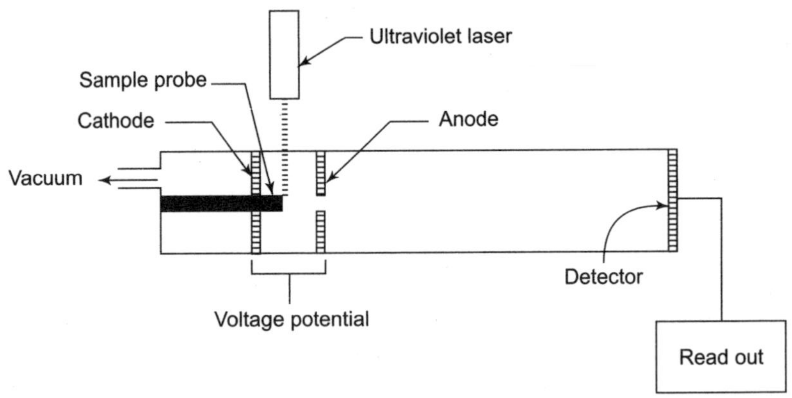

Fig. 6.7 A Schematic Presentation of MALDI

The first step in the MALDI analysis is to dissolve a polymer sample in an appropriate solvent and then a special ingredient like *trans*-cinnamic acid or 2,5-dihydroxybenzoic acid is added . The special ingredient varies from polymer to polymer, but the important thing is that special ingredient has to absorb ultraviolet light. UV absorber is usually about 104 times more than the polymer sample. Once, all is mixed together, the sample is placed in an airtight chamber, on the tip of the sample probe. With a vacuum pump all the air is sucked out of the chamber. The solvent evaporates when air is sucked out and the sample to be analyzed is left with layer of UV-absorbing compound with a little bit of polymer in it i.e., the polymer is now dispersed in a matrix of the UV-absorbing compound. This is why it is called matrix-assisted laser desorption/ionization mass spectrometry. Usually an ultraviolet laser in 330–360 nm range is used. The UV absorbing material in the sample absorbs all the energy it can from the laser and also passes some of the absorbed energy to the polymer molecules. The matrix material also reacts with the polymers in such a way that the polymers become charged ions. Usually polymer molecules are way too big and heavy to evaporate, but after absorbing energy and at high temperatures and low pressures, some of the charged polymer molecules evaporate. That is where the word desorption comes from in the name. The floating charged ions are now in gaseous state. There are two electrodes, a positive cathode, and a negative anode at the end of the chamber. When the polymer forms cations, the positive cathode is placed right behind the sample, and the negative anode in front of the sample and vice-versa. The charged polymers go in the direction of the electrode having opposite charge. Most of the time, there is only one single charge on each polymer molecule. This means that the same electrical force is applied on each polymer molecule when it is being accelerated in the electric field between the two electrodes. Acceleration depends on the mass and force applied as

$$F = ma$$

$$\Rightarrow \qquad a = \frac{F}{m}$$

Since the polymer molecules have different masses, therefore there will be different acceleration. At equal force, the more mass, the slower will be the acceleration. The big heavy polymer molecules will take much longer to get to the detector at the end of the chamber and the small ones will hit the detector, first. All the polymer molecules of the same molecular weight will hit the detector together. They hit completely in order by mass. When they hit the detector, the detector registers a peak. The size of the peak is proportional to the number of molecules that hit at one time. This way a series of peaks is obtained (Fig. 6.8).

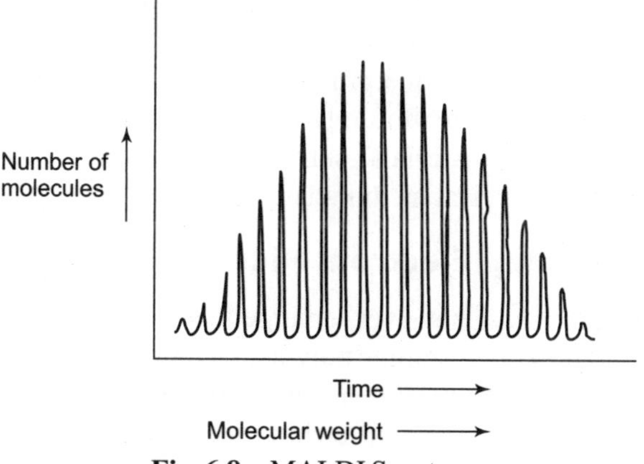

Fig. 6.8 MALDI Spectrum

Since the time a molecule takes to hit the detector is proportional to its mass, what one really has is a plot of molecular weight on the *x*-axis and the number of molecules having given molecular weight on the *y*-axis and the molecular weight distribution is known.

6.5 MORPHOLOGY

6.5.1 X-ray Diffraction Study

6.5.1.1 Wide Angle X-ray Scattering

WAXS is an X-ray diffraction technique that is often used to determine the crystalline structure of polymers. This technique specifically refers to the analysis of Bragg Peaks scattered to wide angles, which (by Bragg's law) implies that they are caused by small structures.

6.5.1.2 Small-Angle X-ray Scattering

SAXS is a small-angle scattering (SAS) technique where the scattering of X-rays (wavelength 0.1 ... 0.2 nm) by a sample which has inhomogeneities in the nm range, is recorded at very low angles (typically < 3–$5°$). In this angular range, information about the shape and size of macromolecules, characteristic distances of partially ordered materials, pore sizes and the like is contained. SAXS is capable of delivering structural information of macromolecules between

5 and 25 nm, of repeat distances in partially ordered systems of up to 150 nm, using laboratory equipment. USAXS (ultra-small angle X-ray scattering) can resolve even larger dimensions.

6.5.1.3 Small Angle Neutron Scattering

SANS is a laboratory technique similar to small angle X-ray scattering. Extremely small features of samples can be studied with elastically scattered neutrons. Small Angle Neutron Scattering (SANS) is an ideal tool for studying the microstructure and macroscopic properties of matter. It occurs whenever inhomogeneities of dimensions in the nanometer scale are present in any condensed matter. The technique provides valuable information over a wide variety of scientific and technological applications including chemical aggregation, defects in materials, surfactants, colloids, alloy segregation, polymers, proteins, biological membranes, viruses, and macromolecules.

6.5.1.4 Scanning Electron Microscopy

Electron microscopy has, however, emerged as a very powerful tool in the morphological studies of crystalline polymers. Direct technique is practiced for structure observation of only thin specimen like polymer single crystals but heavy metal shadowing or solvent etching techniques are used for common polymeric samples.

6.6 THERMAL ANALYSIS

6.6.1 Differential Scanning Calorimetry (DSC)

It is a thermoanalytical technique in which the difference in the amount of heat required to increase the temperature of a sample and the reference are measured as a function of temperature. Both the sample and the reference are maintained at nearly the same temperature throughout the experiment. Generally, the temperature program for a DSC analysis is so designed that the sample holder temperature increases linearly as a function of time. The reference sample should have a well-defined heat capacity over the range of temperatures to be scanned. The basic principle underlying this technique is that, when the sample undergoes a physical transformation such as phase transitions, more (or less) heat will need to flow to it than the reference to maintain both at the same temperature. Whether more or less heat must flow to the sample depends on whether the process is exothermic or endothermic. For example, as a solid sample melts to a liquid it will require more heat flowing to the sample to increase its temperature at the same rate as the reference. This is due to the absorption of heat by the sample as it undergoes the endothermic phase transition from solid to liquid. Likewise, as the sample undergoes exothermic processes (such as crystallization) less heat is required to raise the sample temperature. By observing the difference in heat flow between the sample and reference, differential scanning calorimeters are able to measure the amount of energy absorbed or released during such transitions. DSC may also be used to observe more subtle phase changes, such as glass transitions. DSC is widely used in industrial settings as a quality control instrument due to its applicability in evaluating sample purity and for studying polymer curing.

An alternative technique, which shares much in common with DSC, is *differential thermal analysis* (DTA). In this technique it is the heat flow to the sample and reference that remains the same rather than the temperature. When the sample and reference are heated identically, phase changes and other thermal processes cause a difference in temperature between the sample and reference. Both DSC and DTA provide similar information; DSC is the more widely used of the two techniques.

DSC Curves

The result of a DSC experiment is a heating or cooling curve. This curve can be used to calculate enthalpies of transitions. This is done by integrating the peak corresponding to a given transition. It can be shown that the enthalpy of transition can be expressed using the following equation:

$$\Delta H = KA$$

where ΔH is the enthalpy of transition, K is the calorimetric constant, and A is the area under the curve. The calorimetric constant will vary from instrument to instrument, and can be determined by analyzing a well-characterized sample with known enthalpies of transition.

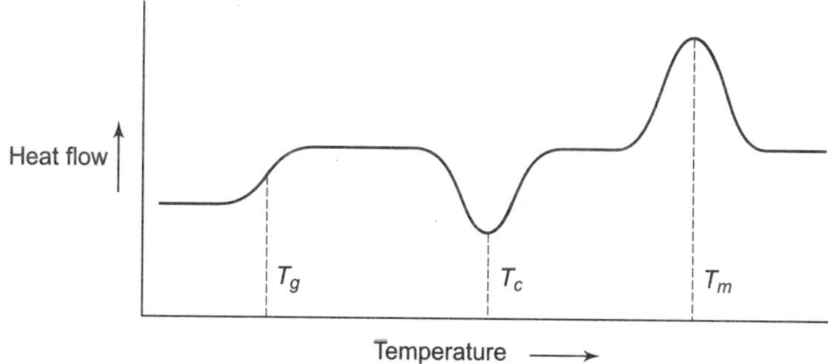

Fig. 6.9 A Schematic DSC Curve Demonstrating the Appearance of Different Transition

Differential scanning calorimetry can be used to measure a number of characteristic parameters of a sample. Using this technique it is possible to observe fusion and crystallization events as well as glass transition temperatures (T_g). In addition to these applications DSC can be used to study oxidation as well as other chemical reactions. Glass transitions occur as the temperature of an amorphous solid is increased. A glass transition is characterized by a decrease in viscosity. These transitions appear as a step in the baseline of the recorded DSC signal. This is due to the sample undergoing a change in heat capacity, but no formal phase change occurs.

As the temperature increases, an amorphous solid will become less viscous. At some point the molecules will obtain enough freedom of motion to spontaneously arrange themselves into a crystalline form. This is known as the crystallization temperature (T_c). This transition from amorphous to the crystalline form is an exothermic process, and results in a peak in the DSC signal. As the temperature increases the sample eventually reaches its melting temperature (T_m). The melting process results in an endothermic peak in the DSC curve. The ability to ascertain transition temperatures and enthalpies make DSC an invaluable tool in producing phase diagrams for various chemical systems.

Of course, not everything which is shown in DSC curve (Fig. 6.4) will be on every DSC plot. The crystallization dip and the melting peak will only show up for polymers that can form crystals. Completely amorphous polymers will not show any crystallization, or any melting either. But polymers with both crystalline and amorphous domains, will show all the features shown above. There is a big difference between the glass transition and the other two thermal transitions, crystallization and melting. For the glass transition, there is no dip, and there is no peak, either. This is because there is no latent heat given off, or absorbed, by the polymer during the glass transition. Both melting and crystallization involve giving off or absorbing heat. The only thing which is shown at the glass transition temperature is a change in the heat capacity of the polymer. Because there is a change in heat capacity, but there is no latent heat involved with the glass transition, the glass transition is called as a second order transition. Transitions like melting and crystallization, which do have latent heats, are called first order transitions.

For the polymer chemist, DSC is a handy tool for studying curing processes, which allows the fine tuning of polymer properties. The cross-linking of polymer molecules that occurs in the curing process is exothermic, resulting in a positive peak in the DSC curve that usually appears soon after the glass transition.

6.7 MECHANICAL PROPERTIES TESTING

Methods of Testing

As the Heat DistortionTemperature indicates the temperature limit, within which, a plastic may be used, a variety of methods are used to determine mechanical performance under a variety of loading conditions.
These may be classified as:

(i) Static (i.e., tensile and shear)
(ii) Transient (i.e. creep and stress relaxation)
(iii) Impact (Izod and Charpy)
(iv) Cyclic (Fatigue test)

Static tests are used to measure the force response when a sample is strained, compressed or shear at a constant rate. These provide a means to characterize the mechanical properties of polymer in terms of modulus, strain and elongation to failure.

Transient tests measure the time response of the force (or stress) on a polymer sample when it is rapidly stretched to a given length (stress relaxation) or the time response of strain when load (stress) is rapidly applied.

Impact test measure the energy required for a sample to fail under different loading histories. Fatigue test determine the number of cycles of applied stress, required for failure.

To measure the tensile strength of a polymer sample, it is stretched with a machine such as an Instron. The sample is stretched after clamping it onto the machine. While the machine is stretching the sample, it measures the amount of force (F) that it is exerted. The stress that the sample is experiencing is calculated by dividing the force by the cross-sectional area (A) of the sample.

$$\frac{F}{A} = \text{stress}$$

The amount of force and stress naturally is increased continuously on the sample until it breaks. The stress needed to break the sample is the tensile strength of the material. Since tensile stress is the force placed on the sample divided by the cross-sectional area of the sample, both tensile stress and tensile strength are measured in units of force divided by units of area, usually N/cm^2. Stress and strength can also be measured in megapascals (MPa) or gigapascals (GPa). Likewise, one can imagine similar tests for compressional or flexural strength. In all the cases, the strength is the stress needed to break the sample. To measure tensile modulus, the amount of stress on sample is increased slowly, and the elongation the sample, undergoes is measured at each stress level. The stress is increased until the sample breaks. Then a plot of stress vs. elongation is drawn as shown below:

The height of the curve, when the sample breaks is the tensile strength, and the tensile modulus is the slope of this plot. If the slope is steep, the sample has a high tensile modulus, which means it resists deformation. If the slope is gentle, then the sample has a low tensile modulus, which means it is easily deformed. There are times when the stress-strain curve is not smooth and straight, like the one shown in Fig. 6.10. For some polymers, especially flexible plastics, we get odd curves (Fig. 6.11).

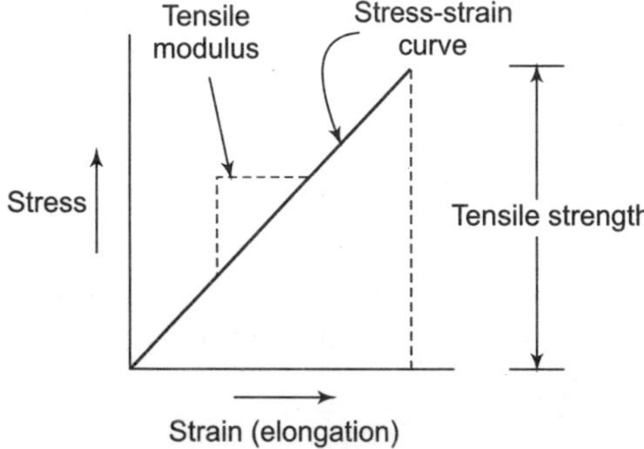

Fig. 6.10 Stress-strain curve

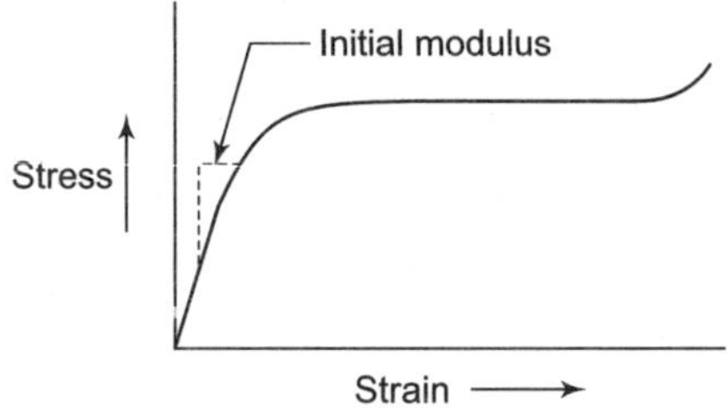

Fig. 6.11 Stress-strain curve for flexible plastics

The slope is not constant as the stress increases. The slope, that is the modulus, is changing with stress. In a case like this, initial slope is usually considered as the modulus. Modulus is measured by calculating stress and dividing by elongation, and would be measured in units of stress divided by units of elongation. But since elongation is dimensionless, modulus is expressed in the same units as strength, such as N/cm^2. In general, fibers have the highest tensile moduli, and elastomers have the lowest, and plastics have tensile moduli somewhere in between fibers and elastomers.

Toughness

The plot of stress vs. strain gives another very valuable piece of information. The area underneath the stress-strain curve is measured as toughness.

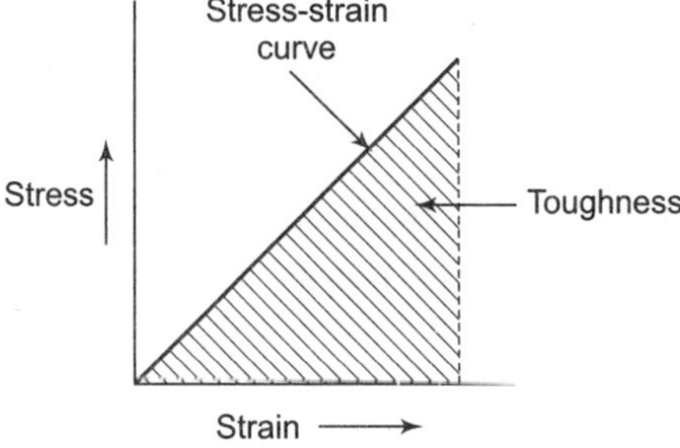

Fig. 6.12 Stress-strain curve to measure toughness

6.7.1 Dynamic Mechanical Analysis or Dynamic Mechanical Thermal Analysis

DMA or DMTA is a technique used to study and characterize materials. It is most useful for observing the viscoelastic nature of polymers. An oscillating force is applied to a sample of material and the resulting displacement of the sample is measured. From this the stiffness of the sample can be determined, and the sample modulus can be calculated. By measuring the time lag in the displacement compared to the applied force it is possible to determine the damping properties of the material.

7

POLYMERIZATION PRACTICES

Learning Objectives

After studying this chapter, you will know:

- About the methods used for polymer synthesis.
- What are the advantages and the disadvatages of various methods?

Chapter Contents

• Bulk polymerization • Solution polymerization • Suspension polymerization • Emulsion polymer • Interfacial polycondensation

Polymerization practice or technique means the specific methods used for synthesis of polymers, how the polymer is prepared or the device used to prepare it. Factors such as (i) nature of monomer, (ii) the type of polymerization mechanism chosen, (iii) the required physical form of the polymer, and (iv) the viability of the process for industrial production are the main criteria of the selection of polymerization technique. Various polymerization techniques which are being used nowadays are discussed here.

7.1 BULK POLYMERIZATION

The **bulk or batch addition polymerization** is the simplest of all polymerization processes. A bulk addition polymerization is a homogeneous system with an organic initiator. There are two possibilities in this type of polymerization:

- If the polymer is insoluble in monomer, then initiation, propagation, and termination might happen in the monomer phase (e.g., polystyrene).

- If the polymer is soluble in the monomer, then the concentration of monomer decreases continuously and the viscosity changes (e.g., vinylidine chloride).

If the polymer is soluble in the monomer, then some physical changes occur with increasing molecular weight (e.g., viscosity). If the polymer is insoluble in the monomer, the rate of initiation is proportional to the monomer concentration, the initiator concentration, and the inverse square of *k* over *t*. The higher the temperature, the lower the molecular weight of the polymer produced. It is thought that the notes here allude to the idea that at higher temperatures, the initiator decomposes to form radicals at a faster rate, then for a given amount of monomer with more radicals present, more polymer chains will be started (initiated), and the resulting polymers will have a lower molecular weight.

In a simple bulk polymerization system, monomer is in liquid state and initiator is dissolved in the monomer. Chain transfer agent may also be dissolved in monomer if molecular weight of polymer is to be controlled. The whole system is, therefore, in a homogeneous phase. The reaction mass is heated or exposed to radiations to initiate the polymerization. As the polymerization proceeds, the viscosity of the solution increases.

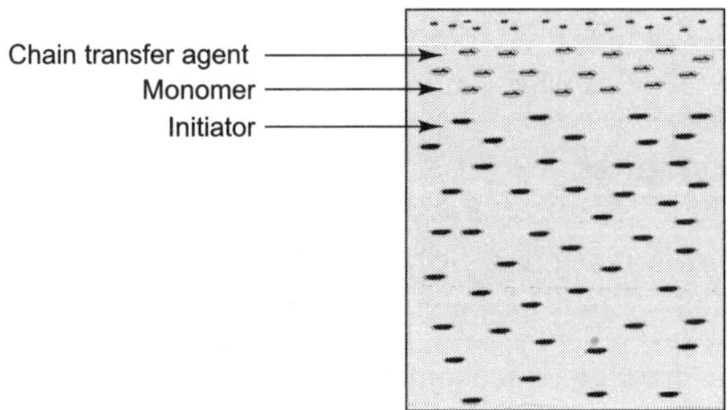

Fig. 7.1 Schematic Representation of Bulk Polymerization Process

Advantages

(i) As only monomer, initiator, and perhaps the chain transfer agent are used, the purest possible polymer may be achieved. This can be important in electric and optical applications.
(ii) It provides the greatest possible polymer yield per reactor volume.
(iii) Polymer recovery is easy.
(iv) It provides options of casting of the polymerization mixture into final product form.

Disadvantages of Bulk Polymerization

(i) Difficulty in removing the last traces of monomer. Removal of unreacted monomer can be a problem. This is a large concern if safe polymer is prepared from monomers which are toxic. The regulations for plastics used in food packaging puts limits on how much monomer can be present in a polymer system used for food containment.

(ii) To keep the bulk polymerization under control, it may have to be run slowly, with the attendant economic disadvantage.

(iii) Bulk polymerization has a built-in hazard i.e., dissipation of heat. The thermal conductivity of monomers and polymers is low, and as the viscosity builds up, the ability for heat transfer via convection is substantially diminished. If the heat energy cannot be dissipated, temperature rises, and at higher temperatures the reaction is going to get faster, so this is a positive feedback loop with disastrous consequences.

Troms-dorff Effect or Gel Effect or Autoacceleration

In bulk polymerization, at lower monomer concentration, the conversion vs. time curve are well described by the equation:

$$\text{In}\,\frac{[M]}{[M_0]} = \frac{k_p}{k_d}\,\frac{f\,k_d[I_0]^{1/2}}{k_t}[e-(k_d^{t/2})-1] \tag{7.1}$$

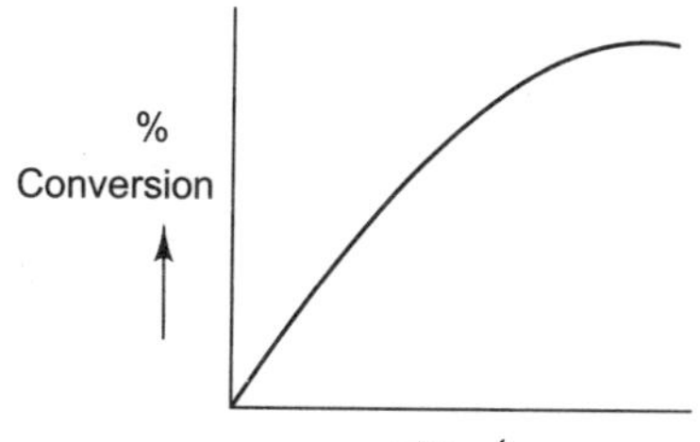

Fig. 7.2 Plot of per cent Polymer Conversion vs. Temperature

At a higher concentration, however, a distinct acceleration of rate of polymerization occurs which does not conform to the classic kinetic scheme. This phenomenon is known as autoacceleration. The reason for this behavior lies in the difference between the propagation and the termination steps and the extremely high viscosity of concentrated polymer solutions.

The propagation reaction involves the approach of a small monomer molecule and a growing chain end, whereas termination requires that the ends of the two growing chains get together. At high polymer concentration it becomes exceedingly difficult for the growing chain lengths to drag them through the entangled mass of dead polymer chains but it is not as difficult for a monomer molecule to pass through the reaction mass owing to its small size. Thus the rate of termination is limited not by the nature of the chemical reaction but by the rate at which the reactants can diffuse them to react, i.e., it is diffusion controlled. This lowers the effective rate constant k_t, and since it appears in the denominator of the equation (7.1), the net effect is to increase the rate of polymerization. The difficulties are compounded by the exothermic nature of final polymerization. The generated heat is not dissipated properly due to high viscosity. This raises the temperature of the reaction mixture, further increasing the rate of reaction.

7.2 SOLUTION POLYMERIZATION

The addition of an inert solvent to a bulk polymerization mass minimizes many of the difficulties encountered in the bulk system. It reduces the tendency towards autoacceleration in free radical polymerization. The main advantage of an inert diluent (either water or an organic solvent) is to take up the heat of polymerization.

For solution polymerizations, there are two possibilities:

- Monomer is soluble and the polymer is soluble in the diluent, e.g., polystyrene in toluene.
- Monomer is soluble and the polymer is insoluble in the diluent, e.g., acrylonitrile in chloroform.

Advantages

(i) The advantage of solution polymerization over bulk polymerization is better heat control.

(ii) Since the reactions are more likely to follow known theoretical kinetics, the design of the reactor system is facilitated.

(iii) For some applications, e.g., lacquers, the desired polymer solution may be obtained directly from the reaction.

Disadvantages

(i) Large amounts of expensive, flammable and perhaps toxic solvents are required.

(ii) The disadvantage in solution polymerization is the removal of the diluent from the polymer. This requires a distillation, and that costs an appreciable amount of money.

(iii) Since the rate and average chain lengths are proportional to the concentration of monomer in free radical addition, the use of solvent lowers them. Additional lowering of average degree of polymerization will occur, if the solvent acts as chain transfer agent.

Ionic polymerizations are almost exclusively solution processes. Most Ziegler-Natta polymerizations are solution processes. Important water soluble polymers that can be prepared in aqueous solution include polyacylic acid, polyacylamide and polyvinyl alcohol. Polystyrene, polymethylmethacrylate, polyvinylchloride can be polymerized in organic solvents.

7.3 INTERFACIAL POLYCONDENSATION

A variation in solution polymerization is called as interfacial polycondensation. In this method, one monomer of a condensation pair is dissolved in a liquid, and the other member of the pair in another liquid. The liquids are insoluble in one another. The polymer is not soluble in any and forms at the interface between them. Also one of the phases generally also contains an agent that reacts with the molecule of condensation to drive the reaction to completion.

Advantages

Reaction proceeds very rapidly at room temperature and atmospheric pressure, in contrast to the long time, at high temperature and pressure associated with polycondensation. Separation of polymer from the interface is easier even during reaction.

Disadvantages

Separation of solvents and unreacted monomers may cost too high.

7.4 SUSPENSION OR PEARL POLYMERIZATION

Improved heat transfer can be obtained by using excellent thermal properties of water through suspension polymerization. If the monomer is insoluble in water, bulk polymerization can be carried out in suspended droplets, i.e., monomer is mechanically dispersed. The water phase becomes the heat transfer medium. Since it is a continuous phase, viscosity changes very little as the monomer converts to polymer, so the heat transfer is very good. In this system, the monomer must be either i) insoluble in water, or ii) only slightly soluble in water, so that when it polymerizes it becomes insoluble in water.

In suspension polymerization, a batch reactor fitted with a mechanical agitator is charged with a water insoluble monomer and initiator. Sometimes, a chain transfer agent may also be added to control the molecular weight in a free radical polymerization. Droplets of monomer containing the initiator and chain transfer agent are formed. They are typically between 10 to 1000 microns in diameter and serve as miniature reactor for the polymerization. A suspension agent, which is added sometimes, is a material that gives a surface activation that keeps droplets from becoming larger. The behavior inside the droplets is very much like that of the bulk polymerization, but since the droplets are tiny, more rapid reaction rates can be tolerated (than would be the case for bulk polymerization) without boiling the monomer.

Advantages

(i) Better heat control of the reaction.

(ii) Separation of polymer is much easier than in solution polymerization.
(iii) Solvent cost and recovery operations are minimal.
(iv) It only applies to free radical process.
(v) Ionic catalysts do not work because they compete with water.
(vi) Agitation is critical because as the viscosity within the bead rises, the reaction rate increases suddenly (Tromsdoiff effect.) This leads to a surge in heat generation which does not usually occur in solution or emulsion polymerization.

Disadvantages

(i) Low yield per reactor volume.
(ii) A somewhat less pure polymer than that from bulk polymerization, since there are bound to be traces of suspending agents absorbed on the particle surface.
(iii) Only a few monomers are water soluble.
(iv) Reactors capital cost are higher than that of the solution polymerization.

Polymers commonly produced by free radical suspension polymerization include styrenic ion-exchange resins, extrusion and injection molding grades of polyvinyl chloride, poly(sty-co-AN) etc.

7.5 EMULSION POLYMERIZATION

The 'ingredients' for an emulsion polymerization include i) a water soluble initiator, ii) a chemical emulsifier, and iii) a monomer that is only slightly soluble in water, or completely insoluble.

A typical reactor for an emulsion polymerization consists of a water soluble initiator, a chain transfer agent and a surfactant such as the sodium salt of a long chain fatty acid in addition to water and monomer. In emulsion polymerization, the monomer is dispersed in aqueous phase not in the form of discrete droplets, but as uniform emulsion. The emulsion is stabilized by surface active substances, protective colloids and also by certain buffers. Surfactants serve the purpose of lowering the surface tension at the monomer-water interphase and facilitate emulsification of the monomer in water.

Owing to their low solubility, the surfactants get fully dissolved or molecularly dispersed only at low concentration. The excess quantity of the surfactants forms molecular aggregates or micelles or domains and an equilibrium is set up between the dissolved surfactant molecules and the aggregate ones.

The highest concentration where all the molecules are in dispersed phase or the concentration beyond which micelle formation is possible is known as the critical micelle concentration (CMC). Emulsifier molecules are made of two parts- i) a long non-polar hydrocarbon chain to which ii) a polar group, such as ^^^COONa; ^^^^SO_3Na etc. is attached.

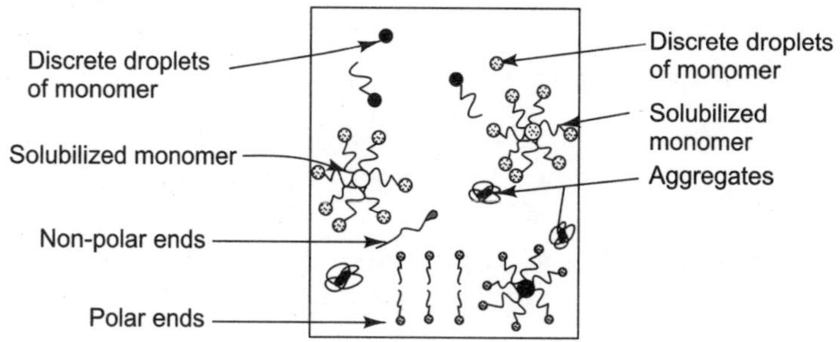

Fig. 7.3 Schematic Representation of Suspension Polymerization Process

In micelle formation, the emulsifier molecules aggregate in such a way that the polar ends of the molecules align themselves outward and the hydrocarbon ends come close to each other at the interior. Due to the close proximity of the hydrocarbon ends of all emulsified molecules, the interior of the micelle acts as a hydrocarbon phase, where the monomer can be solubilized when the monomer is added or agitated, emulsification takes place, the resultant emulsion is a complex system. A molecular solution of the emulsification in water is the continuous phase, where in the monomer droplets and micelles are uniformly dispersed. If the monomer is slightly water soluble, then the aqueous emulsifier solution phase will also contain the part of monomer dissolved in it.

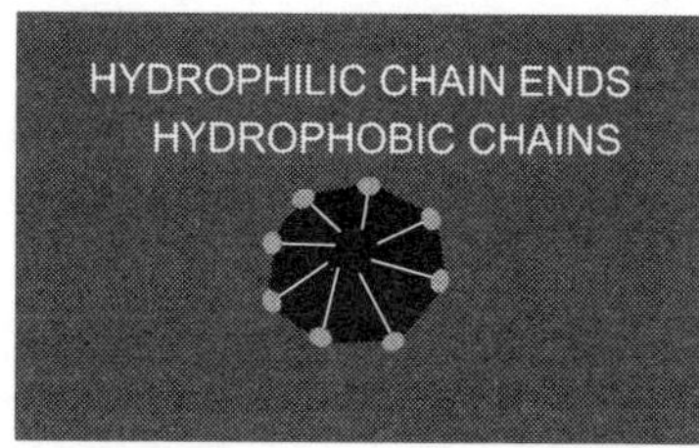

Fig 7.4 Representation of Micelle

Therefore, in emulsion polymerization the monomer is present in the form of discrete droplets and in the solubilized form, in the centre of micelle.

If the initiator is water soluble and monomer is insoluble ,there is hardly any initiator in the monomer droplets, hence no appreciable polymerization occurs inside the droplets. But in the centre of the micelle there is a favorable condition for polymerization because at its hydrophilic end initiator is solubilized and polymerization starts at its surface layer and it proceeds invert, as and when the monomer inside the micelle is consumed. More monomer diffuses into the micelle from the monomer droplets present in the aqueous phase. The chain growth continues till another radical species enters and arrests the chain growth.

Figure 7.5 shows a graph of surface tension as a function of micelle concentration. At a certain concentration, increase in the surface tension results in a sudden increase of the slope of the function.

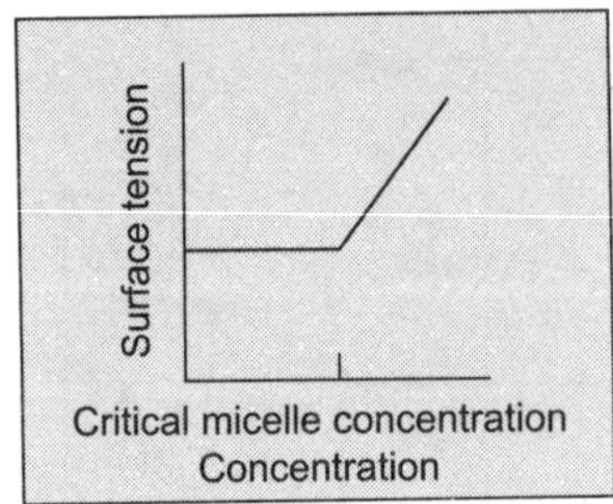

Fig 7.5 Plot of Micelle Concentration vs. Surface Tension

With more and more polymer formation, the polymer chains agglomerates into fine particles and these fine particles are stabilized by the emulsifier layer and dispersed uniformly in the aqueous phase. The end product is usually a stable latexan emulsion of polymer in water rather than a filterable suspension.

Advantages

(1) Ease of control, the viscosity of the reaction mass is much less than that of a true solution of a comparable concentration, the water acts as its heat capacity and the reaction mass may be refluxed.
(2) It is possible to obtain both high rates of polymerization and high average chain length through the use of high surfactant soap and low initiator concentration.
(3) The latex produced is often directly valuable.

Disadvantages

(1) It is difficult to get polymer. The tremendous surface are of the tiny particles providing plenty of rooms for absorbed impurties.
(2) Considerable technology is required to recover the solid polymer.
(3) The water in the reaction mass lowers the yield per reaction.

Kinetics

The free radical kinetics of emulsion polymerization are different from the usual free radical kinetics of bulk, solution and suspension polymerization

$$R_p = [IM_x^o]\ [M]$$
$$R_o = K_p\ [N/Z]\ [M]$$
$$N = \text{number of micelles.}$$

Table 7.1 Differences between Suspension and Emulsion Polymerization

Emulsion	*Suspension*
Initiator must be water soluble and monomer insoluble.	Initiator is monomer soluble.
Monomer may be slightly water soluble or completely insoluble.	Monomer is water insoluble.

8

POLYMER PROCESSING

Learning Objectives

After studying this chapter, you will know:

- How polymers are processed?
- About the various processing methods.

Chapter Contents

- Basic processing operations • Extrusion • Molding • Calendaring
- Casting

8.1 INTRODUCTION

The term processing is used here to describe the technology of converting raw polymer, or compounds containing raw polymer, to articles of a desired shape. Considering the wide variety of polymer types and even wider variety of articles made from them, a complete description and analysis of the myriad of processing techniques that have sprung up through the years would be impossible here. There are several references that devote considerable space to detailed descriptions of polymer processing operations. Similarly, quantitative treatment of processing operations has been the subject matter of a number of books. Here, we outline the common processing techniques, introduce relevant terminology and consider how the various techniques are used in the fundamentals previously discussed.

8.2 BASIC PROCESSING OPERATIONS

Polymer-processing operations may be classified in five broad categories—extrusion, molding, spinning, calendaring and coating. Of there, extrusion is perhaps the most widely used. Applications of extrusion include the continuous production of plastic pipe, sheet, and rods. Molding is normally a batch process, principally in the form of injection—and compression-molding operations used to make plastic parts as diverse as the cap to a ball-point pen to a fibre-reinforced bathtub. Other important processing methods include the fibre-spinning of textiles, coating, the calendaring of plastic sheet, deposition of organic coatings on plastic sheet, molding, stamping and compounding.

8.3 EXTRUSION

Thermoplastic items with a uniform cross-sectional area are formed by extrusion. This includes many familiar items such as pipe, hose and tubing, gaskets, wire and cable insulation, and sheeting. Molding powder is conveyed down an electrically or oil-heated barrel by rotating screw. It melts as it proceeds down the barrel and is forced through a die that gives it its final shape. Vented extruders incorporate a section in which a vacuum is applied to the melt to remove volatiles, such as traces of unreacted monomer, moisture, solvent from the polymerization process, or degradation products.

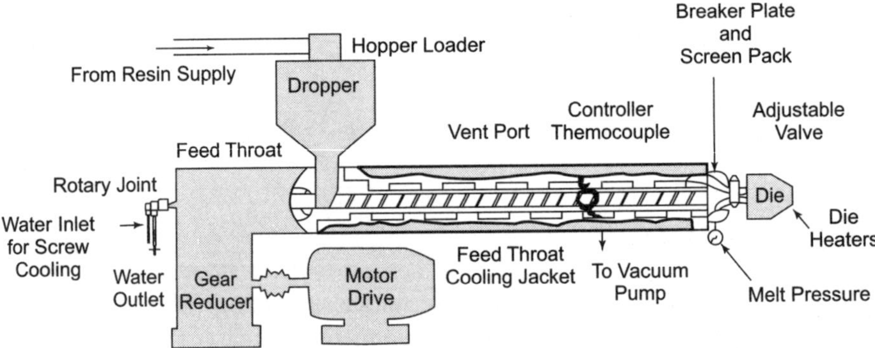

Fig. 8.1 An Extruder

The design of extruder's screws is an interesting and complex technical problem that has received considerable study. Screws are optimized for the particular polymer being extruded. Basically, a screw consists of three sections: melting, compression, and metering. The function of the melting sections is to convey the solid pellets forward from the hopper and convert them into molten polymer. Its analysis involves a combination of fluid and solid mechanics and heat transfer. The compression section, in which the depth of the screw flight decreases, is designed to compact and mix the molten polymer to provide a more or less homogeneous melt to the metering section, the function of which is to pump the molten polymer out through the die. This last section is well understood. Analysis of the metering section is an interesting application of the rheological principles discussed previously, as is much of die design. The determination of the die cross-section needed to produce a desired product, cross section (other than flat or circular) is still pretty much a trial-and-error process, however. Viscoelastic polymer melts well upon emerging from the die (recovering stored elastic energy) and the degree of die swell cannot be reliably predicted.

In addition to die swell, as the extrusion rate is increased, the extrudate begins to exhibit roughness and then an irregular, severely distorted profile comes up. This phenomenon is known as melt fracture. There is not yet general agreement as to its causes, but melt fracture can be minimized by increasing die length, smoothly tapering the entrance to the die, and raising the die temperature.

Extruders are normally specified by screw diameter and length-to-diameter ratio. Diameters range from 1 inch in laboratory or small production machines to a foot for machines used in the final palletizing step of production operations. Typical L/D ratios seems to grow each year or so, with values now in the 20/1 to 36/1 range. Single-screw extruders depend for their pumping ability on the drag flow of material between the rotating screw and stationary barrel. As a result, they are not positive-displacement pumps and tend to give a rather broad residence time distribution. Moreover, they are not particularly good mixing devices. Counter-rotating twin-screw extruders are true positive-displacement pumps, capable of generating the high pressures needed in certain profile extrusion applications. Co-rotating twin-screw extruders, though not positive displacement pumps, with proper screw design, can give excellent mixing and a narrow residence time distribution (subjecting all the material to essentially the same shear and temperature history). They are, therefore, used extensively in polymer compounding (mixing) operations, and to a certain extent a continuous polymerization reactors.

The heaters on extruders are needed mainly for start-up. In steady-state operation, most or all of the heat for plasticizing the polymer is supplied by the drive motor through viscous energy dissipation. In fact, cooling through the barrel walls and/or screw centre is sometimes necessary.

Most packaging film is produced by blow extrusion. A thin-walled hollow cylinder is extruded vertically upward. Air is introduced into the interior of the cylinder, expanding it to a tube of film (less than 0.01 inch thickness). The tube is grasped between rolls at the top, preventing the escape of air and flattening it for subsequent slitting and windup on rolls. The expanded tube is rapidly chilled by a blast of air from a chill ring as if proceeds upward. In the case of polyethylene, this rapid chilling produces smaller crystallites and enhances film clarity.

8.4 CALENDARING

Polymer sheet (greater than 0.01 inch thickness) may be produced by extrusion through thin flat dies or by calendaring Basically, a calendar consists of a series of rotating heated rolls, between which the polymer compound (most often plasticized PVC) is squeezed into sheet form, the thickness of the sheet being determined by the clearance between the rolls. Commercial calendars may be very large (rolls up to 3 ft in diameter by 8 ft. in length) and represent a big capital outlay, but they are capable of tremendous production rates (up to 100 yards/min). In addition to forming unsupported sheeting, the polymer may be laminated to a layer of fabric between two rolls to give a supported sheeting. The final rolls may also be embossed to impart a pattern to the sheet. Shower curtains, vinyl upholstery materials (Naugahyde) and vinyl floor tiles are produced by calendaring.

8.5 SHEET FORMING

Thermoplastic sheet is converted to a wide variety of finished articles by processes known generically as sheet forming. Although the details vary considerably, these processes all involve heating the sheet above its softening point and forcing it to conform to a mold. In vacuum forming, for example, the heat-plasticized sheet (either directly from an extruder or preheated in an oven) is drawn against the mold surface by the application of a vacuum from beneath the surface. Similarly, a positive pressure may be used to force the sheet against the mold surface, and where very deep draws are required, mechanical assists—plug forming – are used. In drape forming, the plasticized sheet is draped over a male mold, perhaps with a vacuum assist.

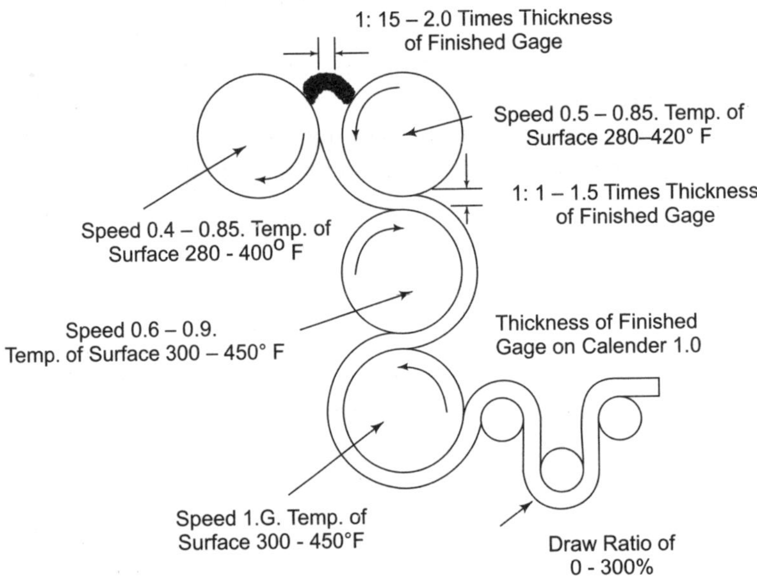

Fig. 8.2 Calender

From a material standpoint, polymers used for sheet forming should have high melt strengths, that is, high melt viscosities so that they do not draw down or thin out excessively or perhaps even tear in the forming operation. Thus high-molecular-weight resins are preferred. One of the great advantages of sheet forming is that relatively inexpensive molds are required since no high pressures are involved. Epoxy molds are often used because they can be easily cast to shape from a hand-made pattern.

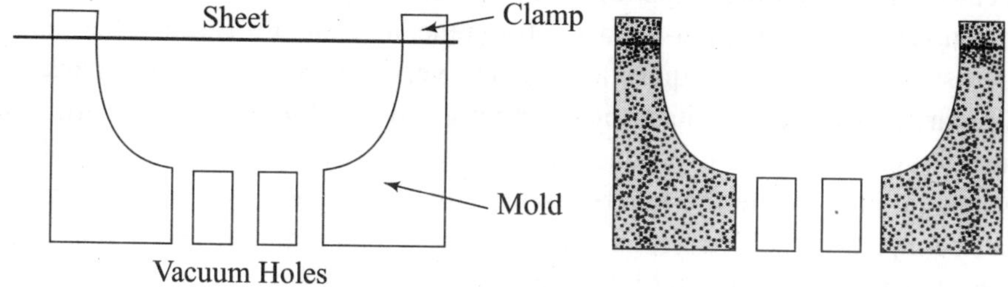

Fig. 8.3 Sheet Forming Process

Among the many familiar items made by sheet forming are drinking cups, meat trays, cigarette packs, aircraft canopies, advertising displays, and lighting globes.

8.6 STAMPING

The plastic industry has long envied the ability to stamp sheet-metal parts with a cycle time of a few seconds. Because of the high production rates possible with a stamping process (this is one important reason why most auto bodies are largely stamped steel), work is progressing with the aim of adapting such a process to forming thin-walled (0.1 inch) parts from glass-reinforced thermoplastic sheet (Azdel). The sheet is heated in an infrared oven and then stamped in modified metal-forming equipment, with a dwell time of about 8 seconds as the material cools in the mold. Because of the thin sections, cooling is rapid and the cycle times as low as 15 to 20 seconds can be achieved with automated feeding and part removal systems. So far, the process has been limited to rather simple shapes and shallow draws (crankcase oil pans, battery trays, bucket seat bottoms and backs, etc.), but because of its economic potential it should achieve more widespread use in the near future.

8.7 CASTING

The raw materials for many thermosetting polymers are available as low-molecular-weight liquids which are converted to solid or rubbery materials by the crosslinking reaction. These materials are easily cast to shape at atmospheric pressure in inexpensive molds. A good example of this process is the production of "carved wood" furniture panels from polyester resins. Only the original need actually be carved (which is the expensive step). A mold is made by pouring a room-temperature curing silicone rubber over the original. When cured, the rubber mold is stripped from the original and used to cast the polyester copies from a liquid unsaturated polyester. A free-radical initiator causes crosslinking to the final solid object, which reproduces faithfully the detail of the carved original at a fraction of the cost.

Such furniture components provide a good example of how the processing method of choice often depends on the production volume. They can also be injection molded from high-impact polystyrene, for example. Because the panels are usually large, they require a large and very expensive molding machine. Similarly, the mold, with its intricate carving and wood-grain detail in hardened steel, is very expensive to produce. Thus the capital outlay for injection-molded panels is huge, while the reverse is true for cast panels. Nevertheless, parts can be injection molded with a cycle time less than a minute and a very little hand labor is involved, and when the capital investment is amortized over a large number of panels, it becomes a small contribution to the total panel cost. On the other hand, the long cycle times and hand labor required for the casting process makes it prohibitively expensive for large production runs, but it is the preferred process for small number of parts.

8.8 SOLUTION CASTING

Plastic sheet or film may be produced by dissolving the polymer in an appropriate solvent, spreading the viscous solution onto a polished surface, and evaporating the solvent. In the manufacture of photographic film base, the solution is spread with a "knife" onto a slowly rotating wheel about 2 ft wide and 20 ft in diameter. As the wheel revolves, heated air evaporates the solvent (which is recovered for reuse), and the dried film is stripped from the wheel before the casting point is again reached.

8.9 BLOW MOLDING

A quick walk through a supermarket provides convincing proof of the economic importance of plastic bottles. They are nearly all made by blow molding. In one form of this process, *extrusion blow molding* (Fig. 8.4), a hollow cylindrical tube or *parison* is extruded downward. The parison is then clamped between halves of a water-cooled mold. The mold pinches off the bottom of the parison

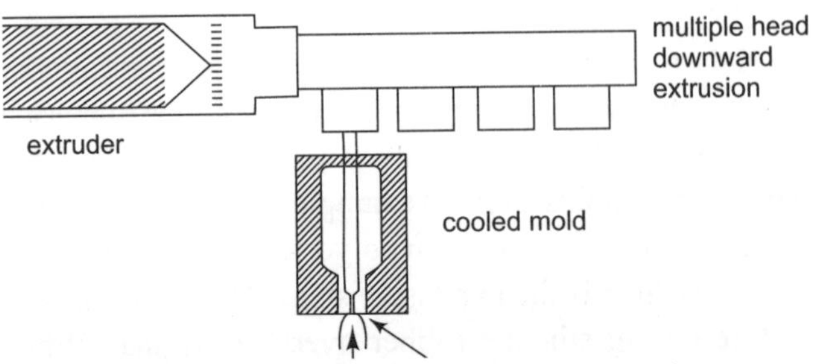

Fig. 8.4 Blow Molding

and forms the threads in the neck of the bottle. Compressed air expands the parison against the inner mold surfaces, and when the part has cooled sufficiently, the mold opens, the part is ejected, and the mold is returned to grab another parison.

The rheological properties of the parison are important. If it sags too much before being grasped by the mold, the walls of the bottle will be too thin in places. Sag is minimized by using high-molecular-weight compounds with high viscosities. Many blow-molding extruders are equipped *with parison programmers* which vary the orifice diameter as the parison is being extruded, providing any desired variation in wall thickness in the blown product.

Extrusion blow molding has passed well beyond the bottle-production stage. It is now being used to produce large items such as drums and truck, automobile, recreational-vehicle and gasoline tanks. In such metal replacement applications, blow-molded containers offer light-weight and great design flexibility.

8.9.1 Injection Blow Molding

In injection blow molding, the parison is formed by injection molding, with the end closed and the threads molded on. A variation on this process known as stretch (or orientation) blow molding is also commercially important. In this process, parisons are injection molded and allowed to cool to room temperature. They are reheated in a radiant-heat oven in which close control is excercised over the temperature profile of the parison. When introduced to the blow mold, the parison is normally a fraction of the length of the mold (Fig. 8.5), but before blowing, a rod rapidly stretches the parison to the full length of the mold, orienting the polymer molecules in the axial direction. This is followed by the blowing process, which imparts tangential (hoop) orientation. The resulting biaxial orientation improves the toughness, creep resistance, clarity, and barrier properties (resistance to permeation) of the bottle material. In this case the frozen-in strains that are normally detrimental to thick injection-molded parts are deliberately introduced to thin-walled blown bottles with highly beneficial results. The biaxial orientation provided by stretch blow molding is absolutely essential to the success of the poly(ethylene terephthalate) bottles that have captured the market.

8.9.2 Reinforced Thermoset Molding

Many plastics, when used by themselves, do not posse enough mechanical strength for structural applications. When reinforced with high-modulus fibers, however, the composites have high strength-to-weight ratios and can be fabricated into a wide variety of complex shapes. Glass and other fibers are used extensively to reinforce thermosetting plastics, mostly polyesters and epoxies. FRPs (fiber-glass-reinforced plastics) are now used for just about all boats under 40 ft in length, truck cabs, low-production-volume automobile bodies, structural panels, aircraft components, and so on. Many such objects are

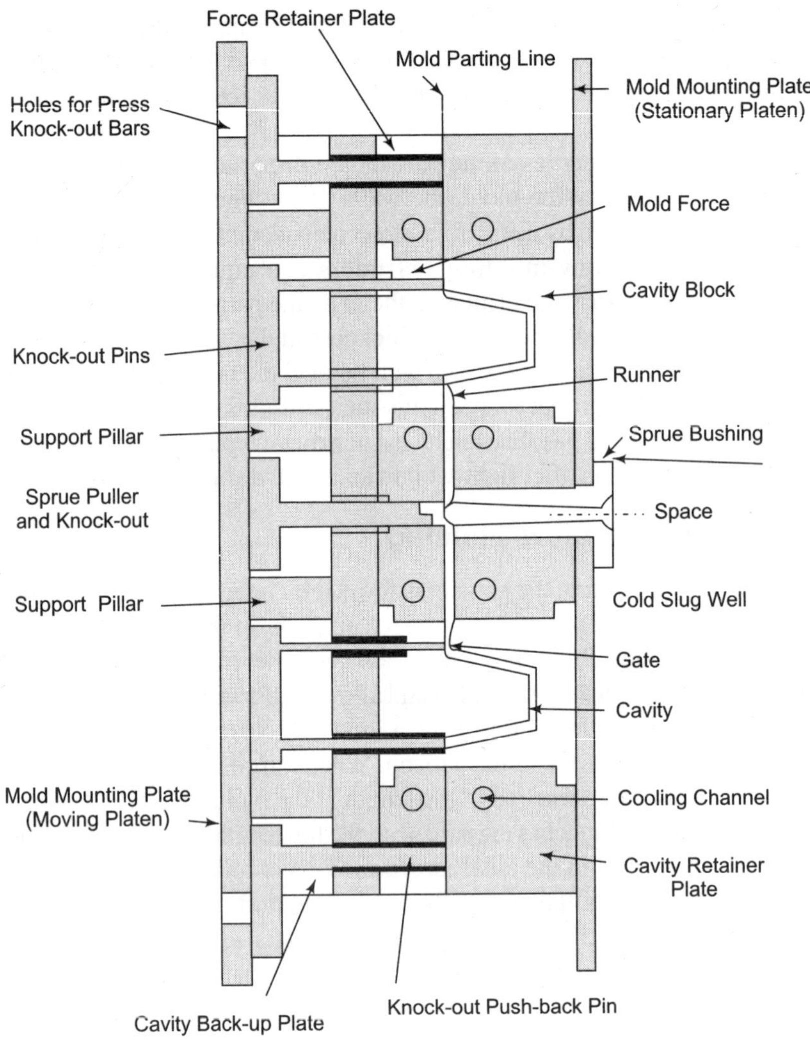

Fig. 8.5 Injection Blow Molding

fabricated by a process known as hand layup. The mold surface is often first sprayed with a pigmented but nonreinforced gel coat of the liquid resin to provide a smooth surface finish. The gel coat is followed up by successive layers of fiber glass, either in the form of woven cloth or random matting, impregnated with the liquid resin, which is then cured (crosslinked) to give the finished product. This process is tremendously versatile. The molds are relatively inexpensive because no pressure is required. Objects may be selectively reinforced by adding extra layers of material where desired. The major drawback of the process is the expense of the hand labor involved, which makes it uneconomical for large production volumes. For this reason it is giving way to a spray-up process. A special gun chops continuous fibers into approximately 1-inch lengths. The chopped fibers are combined with a stream of liquid resin and sprayed directly onto the mold surfaces. Although the random chopped fibers do not reinforce quite as well as woven cloth or random mat, the labor savings are substantial. Sprayed-on polyester-fiber glass backings are also applied to vacuum-formed polymethyl methyacrylate sheets, thus combining a smooth, hard strain- and ultraviolet-resistant acrylic surface skin with the light weight strength of a fiber-glass-reinforced plastic. Such composites are used for sinks, bathtubs, recreational vehicle bodies, etc.

In many objects, stresses are not uniformly distributed; so the reinforcing fibers may be arranged to support the stress most efficiently. In fiber-glass or graphite fishing rods, vaulting poles and golf club shafts, the fibers are arranged along the long axis to resist the bending stresses applied. The process of filament winding extends this principle to more complex structures. Continuous filaments of the reinforcing fiber are impregnated with liquid resin and then wound on a rotating mandrel. The winding pattern is designed to resist most efficiently the anticipated stress distribution. The range of the Polaris submarine ballistic missile was increased by several hundred miles by replacing the metallic rocket casting with a filament-wound reinforced plastic system. This technique is also used to produce tanks and pipe for the chemical process industries, and even gun barrels, by filament winding about a thin metal core that provides the necessary heat and abrasion resistance.

Pultrusion is used to produce continuous lengths of objects with a constant cross section, for example, structural beams. Continuous fibers (roving) and/or mat are impregnated by passing them through a tank of liquid resin and pulled slowly through a heated die of the desired cross section. The resin cures to a solid in the die.

8.10 FIBER SPINNING

The polymers used for synthetic fibers are similar, and in many cases identical to those used as plastics, but for fibers, the processing operation must produce an essentially infinite length-to-diameter ratio. In all cases this is accomplished by forcing the plasticized polymer through a spinnerette, a plate in which a multiplicity of holes have been formed to produce the individual fibers, which are then twisted together to form a thread for subsequent weaving operations. The cross-section of the spinneret holes obviously has a lot to do with the fiber cross section, which in turn greatly influences the properties of the fiber. The three basic types of spinning operations differ mainly in the method of plasticizing and deplasticizing the polymer.

Melt spinning is basically an extrusion process. The polymer is plasticized by melting and pumped through the spinnerette. The fibers are usually solidified by a cross current blast of air as they proceed to the drawing rolls. The drawing step stretches the fibers, orienting the molecules in the direction of stretch and inducing high degrees of crystallinity, a necessity for good fiber properties. Nylons are commonly melt spun.

In dry spinning, a solution of the polymer is forced through the spinnerette. As the fibers proceed downward to the drawing rolls, a counter-current stream of warm air evaporates the solvent. In this process the cross section of the fiber is determined not only by the shape of the spinnerette holes, but also by the complex nature of the diffusion-controlled solvent evaporation process, because there is considerable shrinkage as the solvent evaporates. The acrylic fibers, mainly polyacrylonitrile, are produced by dry spinning.

9

COMMERCIALLY IMPORTANT POLYMERS

Learning Objectives

After studying this chapter, you will know:

- About types of commercially important polymers.
- How are they produced?
- Their properties and uses.

Chapter Details

- Synthetic Thermoplast Organic Polymers
- Synthetic Thermoset Organic Polymers
- Rubber Polymers
- Natural Polymers
- Biodegradable Synthetic Polymers
- Conductive Polymers
- Inorganic Polymers

9.1 SYNTHETIC ORGANIC POLYMERS

9.1.1 Thermoplastics

In this section, a brief description of some important synthetic thermoplastics is given.

Polyethylene

It is the most common and important polymer and is obtained by the polymerization of ethylene.

$$nCH_2{=}CH_2 \xrightarrow{\text{Polymerization}} \begin{array}{cccccccc} H & H & H & H & H & H & H & H \\ | & | & | & | & | & | & | & | \\ C- & C- & C- & C- & C- & C- & C- & C- \\ | & | & | & | & | & | & | & | \\ H & H & H & H & H & H & H & H \end{array}$$

Ethylene Polyethylene

The polymerization of ethylene can be carried out by high pressure method or low pressure method. The polyethylene obtained by high pressure method has a low density (0.920) and so it is also known as high pressure polyethylene or low density polyethylene (LDPE). On the other hand, the polyethylene obtained by low pressure method has a highest density (0.955) and so it is known as high density propylene (HDPE).

The ethylene required for polymerization is obtained either from petroleum fractions or by the dehydration of ethyl alcohol at 360–4000°C using amorphous alumina catalyst. The ethylene used for polymerization must be pure.

$$CH_3CH_2OH \xrightarrow[\text{Alumina}]{360-400°C} CH_2{=}CH_2 + H_2O$$

High Pressure Method In this method high purity ethylene is polymerized at a pressure of 1000–1500 atmospheres and at a temperature of 250–280°C in presence of a free radical catalyst such as oxygen or peroxides. The polymer obtained is low density polyethylene (LDPE), which on cooling becomes solid. It is powdered and stored. The molding polymer is obtained by adding additives (like filters, colouring matter etc) to the molten polymer. The unpolymerized ethylene is recycled.

The high pressure or low density polyethylene is highly branched, though the branching is short. The branching results from the intramolecular chain transfer reactions. The polymer is tough and is chemically inert. Acids and alkalies have no effect on it at ordinary temperature. However ,at elevated temperature it is attacked by both acids and alkalies. It is a poor conductor of electricity. It is used for making films for packings, molded toys and articles of domestic use, pipes and squeeze bottles.

Low Pressure Method In this method pure ethylene is polymerized in a reactor at pressure less than 35 atmospheric pressure and temperture of 50–200°C in presence of catalyst such as titanium tetrachloride in a hexane solution of aluminium triethyl (Zeiglar-Natta catalyst). Cyclo hexane acts for dissipating the heat of the reaction and controls consumption of ethylene. The polymer is obtained by centrifugation. The high density polyethylene obtained by this method are linear molecules having density (0.05–0.97). These are chemically inert, possess greater tensile strength and high softening temperature (135°C). It is used for making polyethylene coated papers, flexible bottles and pipes. It is prepared from propylene (the IUPAC name is propene) which comes from the cracking process.

$$\underset{\text{Propylene}}{\begin{array}{cc} H & H \\ | & | \\ C = & C \\ | & | \\ H & CH_3 \end{array}} \xrightarrow[\text{Process}]{\text{Zeigler}} \underset{\text{poly(propylene)}}{-[-\begin{array}{cc} H & H \\ | & | \\ C - & C \\ | & | \\ H & CH_3 \end{array}-]_n-}$$

Propylene is cheaper than ethylene because of its stability factor. The free radical process is not used for the production of polypropylene from propylene because of the extensive transfer of hydrogens to the propagating centers which results in a resonance stabilized alkyl radical which has little tendency to react with another monomer molecule. Advantages of the Ziegler process include better control of the *tacticity*. The product isolated is mostly isotactic.

The physical properties of polypropylene resemble those of high density polyethylene (HDPE). It has increased softening point. It is more prone to attack by oxidizing agents on the tertiary hydrogen. It has higher melting point (165 °C) than the rest of hydrocarbon polymers. It shows high degree of crystallinity. Solubility is similar to PEs. It has better weather aging (this includes exposure to water, sunlight, and corrosive salts etc).

Applications include the plastic bottle industry and injection molding transportation film, automotive parts and fiber formation.

Polystyrene

It is obtained by the polymerization of styrene solution in ethyl benzene in presence of benzoyl peroxide (free radical initiator) by solution polymerization method.

$$\underset{\text{Styrene}}{C_6H_5-CH{=}CH_2} \xrightarrow{\text{Polymerization}} \underset{\text{Polystyrene}}{\left[\begin{array}{cc} H & H \\ | & | \\ -C- & C- \\ | & | \\ C_6H_5 & H \end{array} \right]_n}$$

The styrene required is obtained by the catalytic reaction of ethylene with benzene to give ethyl benzene followed by dehydrogenation.

$$\underset{\text{Ethylene}}{CH{=}CH_2} \quad \underset{\text{Benzene}}{C_6H_6} \xrightarrow{\text{catalyst}} \underset{\text{Ethyl benzene}}{C_6H_5-CH_2H_3} \xrightarrow[\text{Al}_2\text{COH}_2]{600°} \underset{\text{Styrene}}{C_6H_5-CH=CH_2}$$

Polystyrene is a transparent, light (5p. Gr. 1.05 to 1.07 g/cm^3), stable and moisture resistant. It is not affected by acids or other chemicals. However it has low softening range (90–100°C) and is brittle. It is used in making toys, combs, buttons, buckets, containers, radio, television and refrigerator parts. It is also used in making high frequency electric insulators and lenses.

Acrylic polymers

They are based on acrylic acid and its homologues and their derivatives. The principle commercial acrylic polymers are polymerized from the following monomers:

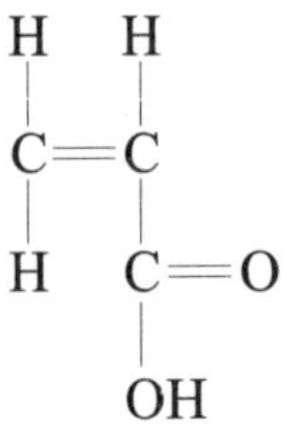

```
H   H            H   CHs            H   H
|   |            |   |              |   |
C = C            C = C              C = C
|   |            |   |              |   |
H   C = O        H   C = O          H   C = O
    |                |                  |
    OH               OH                 O — R
```

Acrylic acid 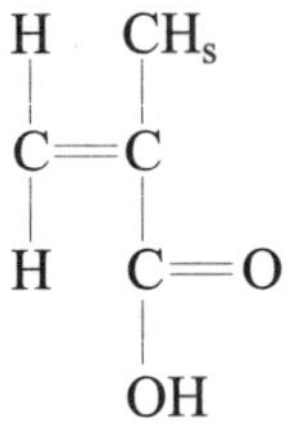 — Methacrylic acid 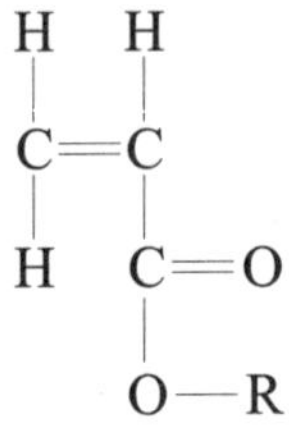 — Esters of acrylic acid

R is a nonhydrogen group.

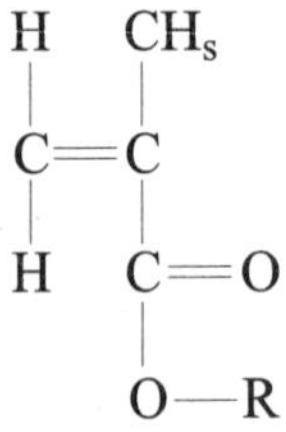

```
H   CHs          H   H              H   H
|   |            |   |              |   |
C = C            C = C              C = C
|   |            |   |              |   |
H   C = O        H   C              H   C = O
    |                |||                |
    O — R            N                  NH2
```

Esters of methacrylic acid 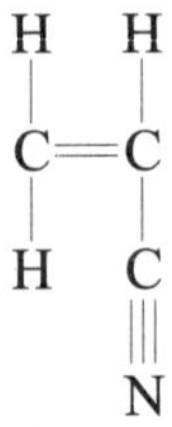 — Acrylonitrile 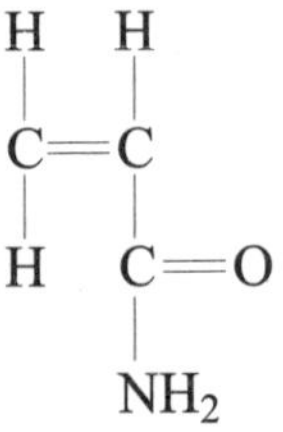 — Acrylamide

The above suggests that acrylamide and acrylonitrile are in the acrylic family. The polymer encyclopedia gives separate coverage for acrylamide and acrylonitrile. Perhaps it is more appropriate to say that acrylamide and acrylonitrile are similar to acrylics, than to say they are acrylics. Applications of acrylics include coatings, textiles, plastics, and plexiglass.

Poly(acrylic acid)

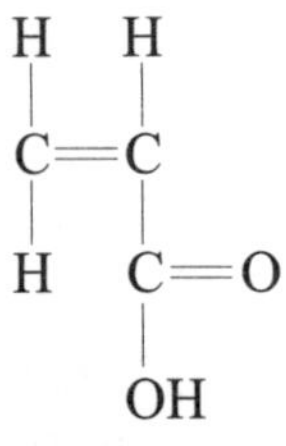

```
H   H                H   H
|   |                |   |
C = C              [ C – C –]n-
|   |                |   |
H   C = O            H   C = O
    |                    |
    OH                   OH
```

Acrylic acid 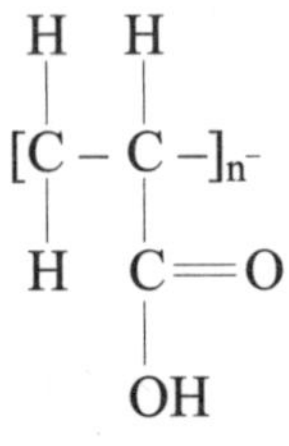 — Poly(acrylic acid)

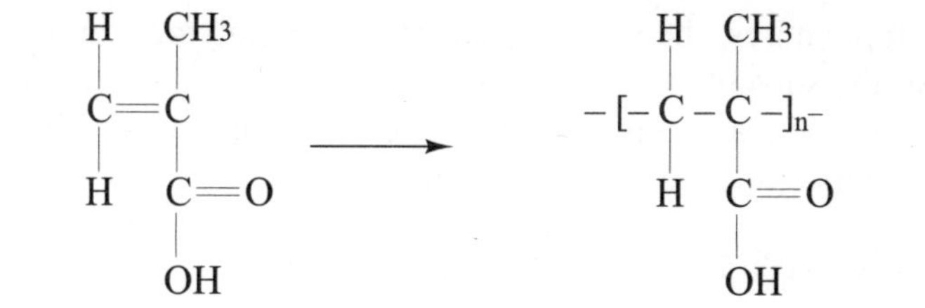

```
H   CH3                          H   CH3
|   |                            |   |
C = C          ——————>      -[-  C - C  -]n-
|   |                            |   |
H   C = O                        H   C = O
    |                                |
    OH                               OH
```

Methacrylic acid → Poly(methacrylic acid)

Polyacrylic acids are synthesized by a free radical process using solution polymerization and water is generally used as the solvent. They are used in water treatment processes and their gels are used for retarded drug delivery.

Poly(methylmethacrylate)

PMMA can be polymerized by bulk, solution, suspension and emulsion polymerization techniques.

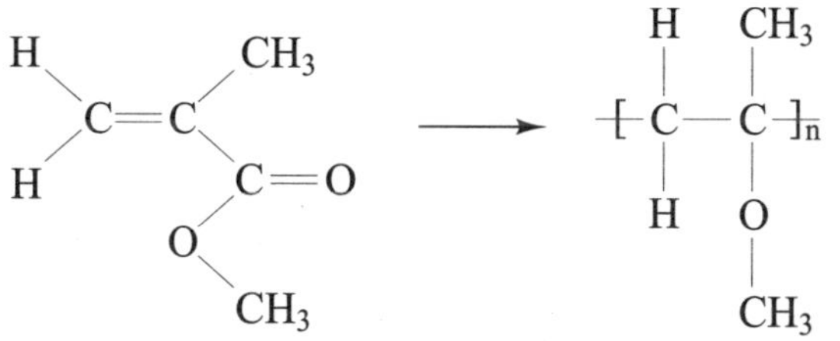

```
H            CH3                    H    CH3
  \         /                       |    |
    C == C         ——————>      -[- C —— C -]n
  /         \                       |    |
H            C == O                 H    O
            /                            |
          O                              CH3
            \
             CH3
```

MMA → PMMA

It is hard, rigid, transparent (very clear to see through). Its softening point is at 125°C. It is tougher than polystyrene but less tough than ABS (acrylobutylstyrene) polymer. It absorbs very little visible light but there is a 4 per cent reflection at each polymer-air interface for normal incident light. PMMA is a polar material and has a rather high dielectric constant. It is a good electrical insulator at low frequencies but less satisfactory at higher frequencies. It has good water resistance. PMMA prepared by free radical polymerization is amorphous and is therefore easily soluble in solvents with similar solubility parameters such as benzene, toluene, chloroform, methylene chloride, esters, ethyl acetate, and amyl acetate. PMMA has good resistance to alkalis (sodium hydroxide etc), aqueous inorganic salts the Pacific and dilute acids. PMMA has a better resistance to hydrolysis than PMA probably by virtue of the shielding of the methyl group. PMMA's outstanding good outdoor weather resistance is marketably superior to other thermoplastics. When heated at about 200°C, decomposition becomes appreciable and at 350 – 450°C, a nearly quantitative yield of monomer is readily obtained. Thus, the recovery of monomer from scrap is feasible Because it is a thermoplastic, it can be molten and molded (at 100 to 150°C) into anything we want. A syndiotactic polymer can be polymerized.

At lower temperatures, the stereochemistry of the polymer can be controlled by means of the solvent.

Applications include automobile lacquers, electrical insulation, good outdoor weather resistance and transparent sheets.

Poly(acrylonitrile)

$$\underset{\text{Acrylonitrile}}{H_2C{=}CH{-}C{\equiv}N} \xrightarrow[\text{polymerization}]{\text{free radical / anionic}} \underset{\text{Polyacrylonitrile}}{\left[CH_2{-}CH(C{\equiv}N) \right]_n}$$

PAN can be polymerized from acrylonitrile using a suspension process. PAN can also be polymerized in aqueous solutions of concentrated inorganic salts such as calcium thiocyanate, sodium perchlorate, and zinc chloride. Emulsion polymerization of PAN is problematic because the monomer has appreciable water solubility.

PAN is toxic only if burned. When PAN is solution polymerized in organic solvent, the solution can be used for fiber spinning. Suitable organic solvents include dimethylacetamide, DMF, DMSO. Fibers prepared from pure PAN are difficult to dye, so a minor portion (~10%) of comonomers such as methyl methacrylate, vinyl acetate, or N-vinyl pyridine are added to improve its dye ability. PAN is extremely soluble in polar solvents such as THF and DMSO. In PAN, appreciable electrostatic forces occur between the dipoles of adjacent nitrile groups on the same polymer molecule. This intramolecular interaction restricts bond rotation and leads to a stiff chain. As a result, PAN has a high crystalline melting point (T_m) of 317 °C and is soluble only in the solvents mentioned above.

Homopolymers of polyacrylonitrile have been used as fibers in hot gas filtration systems, outdoor awnings, sails for yachts, and even fiber reinforced concrete. But mostly copolymers containing polyacrylonitrile are used as fibers to make knitted clothing, like socks and sweaters, as well as outdoor products like tents. Most of the acrylic clothing is made out of some copolymer of polyacrylonitrile either with methyl acrylate or with methyl methacrylate. Copolymers of acrylonitrile and vinyl chloride are flame-retardant, and the fibers made from them are called modacrylic fibers. Poly(styrene-co-acrylonitrile) (SAN) and poly(acrylonitrile-co-butadiene-co-styrene) (ABS), are used as plastics.

Poly(acrylamide)

PAM is water soluble to infinite molecular weight. PAM is prepared by a free radical reaction, but it can also be prepared via alkoxide (RO-) initiation.

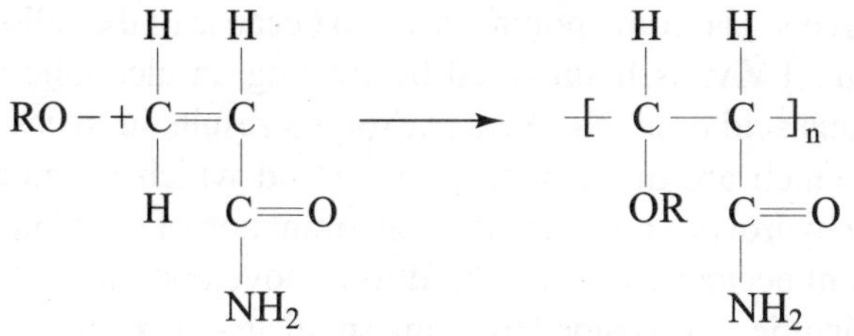

$$\mathrm{RO^- + \underset{H}{\overset{H}{C}}{=}\underset{\underset{NH_2}{C=O}}{\overset{H}{C}} \longrightarrow -\!\![\,\underset{OR}{\overset{H}{C}}-\underset{\underset{NH_2}{C=O}}{\overset{H}{C}}\,]_n-}$$

(second carbon has a negative charge)

PAM is hard and brittle. It is readibly soluble in cold water, and is slightly soluble in organic compounds because of its polarity. It undergoes reactions characteristic of the amide group.

PAM is used as a flocculant in the processing of minerals, the treatment of industrial wastes. Copolymers which incorporate acrylamide increase the dry strength of paper.

Poly(vinyl acetate)

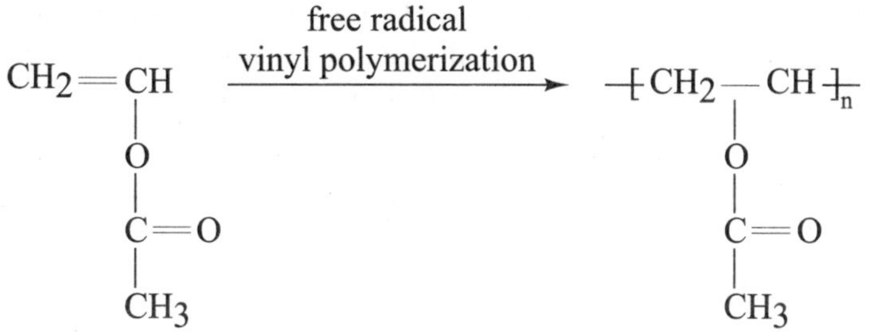

$$\underset{\text{vinyl acetate}}{\mathrm{CH_2{=}CH{-}O{-}C({=}O){-}CH_3}} \xrightarrow[\text{vinyl polymerization}]{\text{free radical}} \underset{\text{poly(vinyl acetate)}}{\mathrm{-\!\![CH_2-CH(O{-}C({=}O){-}CH_3)]_n-}}$$

Poly(vinyl acetate) may be polymerized by bulk, solution, suspension, or emulsion techniques. Commercially, emulsion polymerization is the predominant method. The resulting latex is used "as it is" (as opposed to decanting the water and then drying the solid polymer.) The solid polymer is not isolated, because the end use is for surface coatings and adhesives. PVAc is soluble at room temperature in a wide range of solvents, e.g., benzene, toluene, carbon tetrachloride, methylene chloride, methanol, ethyl acetate, butyl acetate, acetone, MIB ketone.

PVA is used to make wood glues, as well as other adhesives. Paper and textiles often have coatings made of PVA and other ingredients to make them shiny.

Poly (vinyl alcohol)

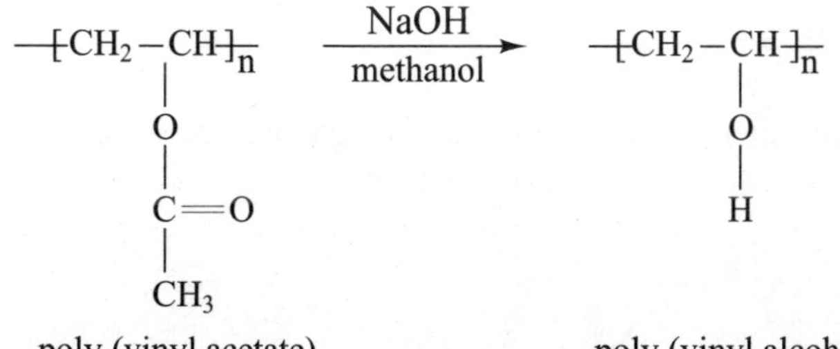

$$\underset{\text{poly (vinyl acetate)}}{\mathrm{-\!\![CH_2-CH(O{-}C({=}O){-}CH_3)]_n-}} \xrightarrow[\text{methanol}]{\text{NaOH}} \underset{\text{poly (vinyl alcohol)}}{\mathrm{-\!\![CH_2-CH(O{-}H)]_n-}}$$

PVA is not prepared via addition polymerization because of the following keto-enol tautomerism. PVAc is hydrolyzed by treating an alcoholic solution of PVAc with aqueous acid or alkali. Acid hydrolysis results in traces of the acid in the polymer which are difficult to remove and which promote polymer instability. Alkali hydrolysis results in the contamination of the product by large amounts of sodium acetate which are hard to remove and have little intrinsic value. These difficulties are avoided by using small amounts of base as catalyst. Poly (vinyl alcohol) is soluble in highly polar solvents such as DMSO, DMF, and water.

Poly(vinyl chloride)

The preparation of poly(vinyl chloride) is mainly done by suspension polymerization, carried out batch wise in a stirred reactor, jacketed for heating and cooling. The reaction is run at 50 to 100 °C and the pressure must be 100 psi. This reaction requires monomer, water, surfactant, and stirring of the initiator. The shape of the polymer particles depends on the suspending agent used. The syndiotacticity of PVC may be increased by polymerizing below –40 °C, using highly active In (indium) and the result is a more brittle product. Atactic PVC has better qualities for commercial applications.

$$H_2C{=}CHCl \longrightarrow \left[-CH_2-CHCl-\right]_n$$

Vinyl chloride → Poly(vinylchloride)

PVC is colorless and rigid. The high chlorine content of PVC renders it flame retarding property. The glass transition temperature of 50–60°C precludes the use of PVC for hot water pipes (the hot water would take the polymer into the rubbery transition and over a period of time there could be dimensional distortion). PVC is soluble in proton acceptor solvents such as THF, esters, and MEK. PVC is unaffected by mild acids, basic materials and water. Exposure to temperatures above 70 °C and exposure to UV light has adverse effects on its properties. PVC is attacked by strong oxdizing agents. PVC can be plasticized with DSP aryl phosphates (tritolyl phosphate), esters of aliphatic acids (dibuytl sebocate), epoxidized oil (soyabean oil), 40–60 parts of plasticizer per 100 parts of polymer.

Fluoropolymers

Most important commercial fluoropolymers nowadays are the homopolymers of:

$$\underset{\text{Tetrafluoroethylene}}{\mathrm{CF_2{=}CF_2}} \qquad \underset{\text{Chlorotrifluoroethylene}}{\mathrm{CF_2{=}CFCl}}$$

$$\underset{\text{Vinyl fluoride}}{\mathrm{CH_2{=}CHF}} \qquad \underset{\text{Vinylidene fluoride}}{\mathrm{CH_2{=}CF_2}} \qquad \underset{\text{Vinylidene fluoride-hexafluoropropylene}}{\mathrm{CF_2{=}CF(CF_3)}}$$

They show excellent thermal (300 to 350 °C) and chemical resistance. These properties result from the great stability of the C=F bond, and also the small size of the fluorine, which makes it possible for the fluorine atoms to pack tightly about the polymer backbone. They are self-lubricating and opaque.

Poly(tetrafluoroethylene)

It is prepared via suspension polymerization. The conventional suspension polymerization techniques produce a granular product suitable for molding and extrusion. When it is polymerized under pressure in stainless steel autoclaves with free radical initiators such as ammonium persulfate, the reaction is rapid and exothermic and requires careful temperature control. The product is a high molecular weight polymer.

It is more commonly known by its trade name, teflon, is incorporated with other materials to induce the self-lubricating effect. Teflon also exhibits the property known as cold flow. At room temperature, teflon changes shape when pressure is applied. Note that in principle, this is true of any polymer, but for teflon, the time scale is short enough for the effect to be appreciable on time scales that are significant to us.

$$\underset{\text{Tetrafluoroethylene}}{\mathrm{F_2C{=}CF_2}} \longrightarrow \underset{\text{Teflon}}{\mathrm{{+}CF_2{-}CF_2{+}_n}}$$

Teflon is a white opaque solid with a waxy appearance and feel. It is a tough flexible material of moderate tensile strength with a tendency to creep under compression (this is an inference to cold flow). Teflon has outstanding electrical insulating properties, and of course, its non-stick properties are excellent. Teflon has a high crystallinity, and a chemical resistance to everything except molten alkali metals and fluorine. It is insoluble in everything except fluorocarbon oils at very high temperatures (as high as the crystalline melting point temperature). It has a good thermal stability and weather resistance. At the crystalline melting point, there is an increase in volume of 25 per cent in the polymer because of a weak translucent material which has a high viscosity, and will not flow. Teflon is self lubricating and its non-stick properties correspond to its very low coefficient of friction. The relatively stiff coil makes teflon slippery. Teflon has a low glass transition temperature. It can not be plasticized because the plasticizer will leach out.

Applications: wire coatings, gaskets, holders and spacers, pump parts, lab equipments, non-stick coverings in kitchen utensils, low friction linings for hoppers, diaphragms for low temperature equipments.

Polychlorotrifluoroethylene

It is prepared by suspension polymerization using procedures similar to those used for teflon. Emulsion polymerization is also used to obtain polymer of a high molecular weight. Reaction temperatures are kept in the range of 0 to 40 °C. Initiation is carried out by means of redox systems activated by a metal salt such as persulfate-bisulphite with a ferrous salt.

It is a clear solid. It has a greater tensile strength than Teflon. It has a greater hardness and resistance to creep than Teflon. It has a lower softening point due to the chlorine atom. It has a lower tendency to crystallize. The presence of the chorine atom produces a greater interchain attraction and accounts for the better mechanical strength. The unsymmetrical structure leads to poor electrical insulation properties. Since there are no hydrogens, there is more resistance to oxidation. There is a good chemical resistance (not quite as good as Teflon).

Poly(vinyl fluoride)

The product may be obtained by polymerization of vinyl fluoride in the presence of water, using benzoyl peroxide as the initiator at 80 °C and pressures up to 1000 atm. PVF has properties resembling the chain properties of PVC, but its molecular weight properties are generally superior because of its greater ability to crystallize. Applications include glazing and pecking.

Poly(vinylidene fluoride)

Poly(vinylidene fluoride) is made from the polymerization of vinylidene fluoride in an aqueous medium under pressure. The polymer may be melt processed by the standard techniques of injection molding and extrusion.

Poly(amides) and Related Polymers

A polyamide contains the amide group [—NH—C=O—] in the main polymer chain. Nylons are the most common and important commercial examples of the synthetic polyamides which are capable of fiber formation. Fatty polyamides, the more complex synthetic polyamides which are not fiber-forming, are used in coatings and adhesives.

Nylons

Nylons can be prepared from:

- reaction of a diamine with a dicarboxylic acid
- the self condensation of a ω-amino acid
- the ring opening polymerization of a lactam

Nylons are distinguished based on the number of carbon atoms in the starting material. Aromatic polyamides like Kevlar and Nomex are made of terephthalic acid and diamino-benzene. Fatty polyamides are obtained by the reaction of di- and polyfunctional amines with polybasic acids which in turn were prepared from unsaturated seven member ring structure shown below—caprolactam. It has six carbon atoms and is called Nylon 6. Similarly, Nylon 66, Nylon 6, 11 are the names given on the number of carbon atoms present in their starting materials. Nylon 66, the largest produced nylon, is prepared using adipic acid. Nylon 6, the second largest produced, is prepared from caprolactam.

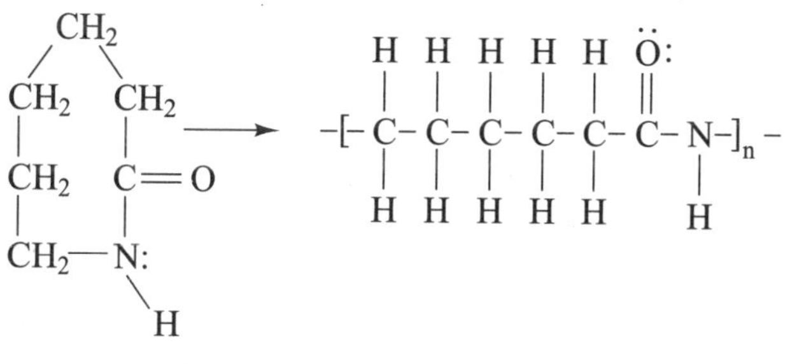

```
      CH2
     /   \
  CH2     CH2                H  H  H  H  H  Ö:
   |       |                 |  |  |  |  |  ||
   |       |      ——→     -[-C- C- C- C- C- C-N-]n -
  CH2     C=O                |  |  |  |  |     |
   |       |                 H  H  H  H  H     H
  CH2 ——  N:
           \
            H
```

Nylons have high impact strength, toughness, flexibility and abrasion resistance. Crystallinity readily occurs and gives materials of high melting point and tensile strength. The mechanical properties are considerably affected by the amount of crystallinity. Increasing the length of the aliphatic segment reduces the melting point, the tensile strength, the heat distortion temperature, and the water absorption, but increases the impact strength and elongation. Textile fibers are usually prepared from Nylon 66 or Nylon 6 because these polymers have the highest tensile strengths. Brushes, sports equipment and surgical sutures are normally Nylon 6, Nylon 10 or Nylon 11, which have greater flexibility and water resistance. Nylon fibers have satisfactory electrical insulating ability under low frequencies and dry conditions. Nylon fibers are soluble in acetic acid, formic acid, and phenols. The oxidation of nylons occurs in air exposure to UV light and or temperature above 70 °C. This oxidation leads to a distortion in mechanical properties.

Nylon 66

Nylon 66 is made from the monomers adipoyl chloride and hexamethylene diamine.

adipoyl chloride + hexamethylene diamine

$$ClOC(CH_2)_4COCl + H_2N(CH_2)_6NH_2 \longrightarrow \left[CO(CH_2)_4CO{-}NH(CH_2)_6NH \right]$$

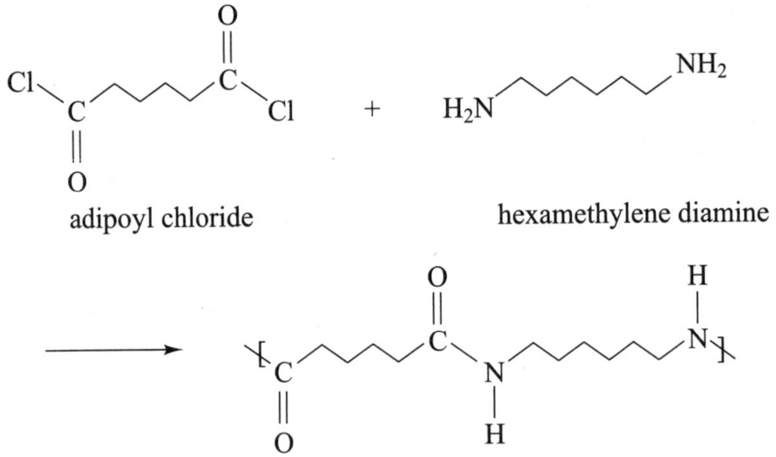

Nylon 66

This is one way of making Nylon 66 in the laboratory. But in a Nylon plant, it's usually made by reacting adipic acid with hexamethylene diamine:

adipic acid + hexamethylene diamine

$$HOOC(CH_2)_4COOH + H_2N(CH_2)_6NH_2 \longrightarrow \left[CO(CH_2)_4CO{-}NH(CH_2)_6NH \right]$$

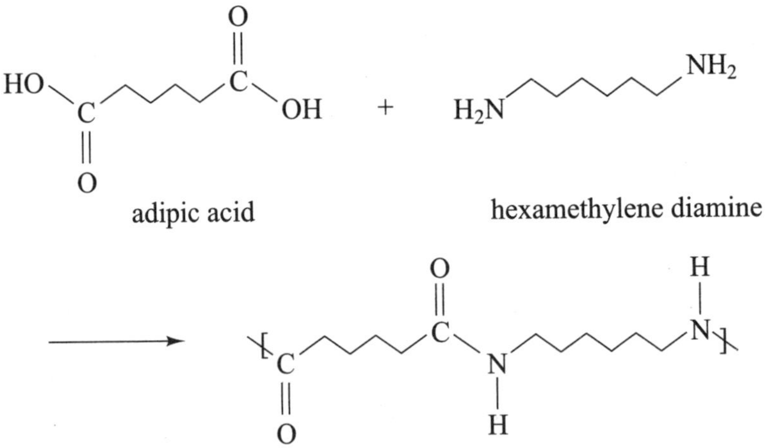

Nylon 66

Nylon 66 has high tanacity and elasticity and is not affected by water. It was remarkable resistance to abrasion. It is used in making dress material, reinforcement of rubber types, manufacture of parachutes, ropes, safety belts, fishing nets, artificial fur coats, undergarments etc.

Nylon 6

It is a lot like Nylon 66 except that it only has one kind of carbon chain, which is six atoms long.

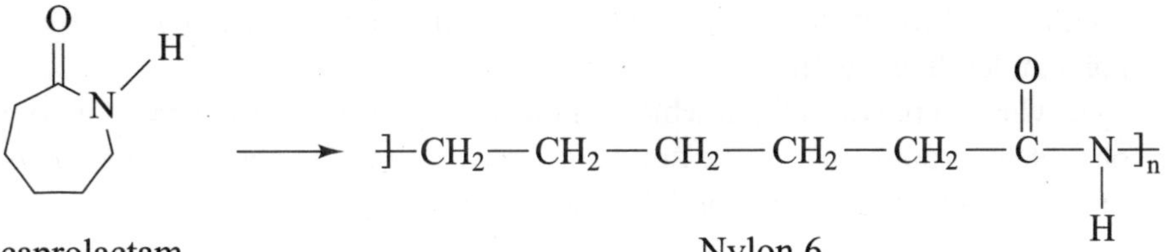

$$\varepsilon\text{-caprolactam} \longrightarrow \left[CH_2-CH_2-CH_2-CH_2-CH_2-\overset{O}{\overset{\|}{C}}-\underset{H}{\underset{|}{N}} \right]_n$$

ε-caprolactam Nylon 6

It is made by a ring opening polymerization from the monomer caprolactam. Nylon 6 does not behave much differently from Nylon 66.

Nylon 610

It is obtained by the condensetion polymerization of hexamethylene diamine (C_6 monomer) and sebacic acid (C_{10} monomer).

$$H_2N-(CH_2)_6-NH_2 \quad + \quad HOOC-(CH_2)_8-COOH$$

Hexamethylene diamine Sebacic acid

$$\downarrow$$

$$-NH-(CH_2)_6-NH-CO-(CH_2)_6-CO-$$

Nylon 610

Nylon 11

It is made by the self concansation of ω-aminoundecanoic acid.

$$H_2N-(CH_2)_{10}-COOH \xrightarrow{\text{Polymerization}} \left[NH-(CH_2)_{10}-\overset{O}{\overset{\|}{C}} \right]_n$$

ω–Aminoundecanoic acid Nylon 11

Nylon 11 along with Nylon 6 is used for molding purpose for gears, bearings, electrical mountings, etc. The nylon bearings last longer and do not require any lubrication.

Aramids

Aramids are aromatic nylons. Aramids include Nomex and Kevlar. Kevlar is used to make things like bullet proof vests and puncture resistant bicycle tyres. Blends of Nomex and Kevlar are used to make fireproof clothing. Aramids are used in the form of fibers. They form into even better fibers than non-aromatic polyamides, like Nylon 66.

Kevlar is a polyamide, in which all the amide groups are separated by *para*-phenylene groups, that is, the amide groups attach to the phenyl rings opposite to each other, at carbons 1 and 4 as shown below:

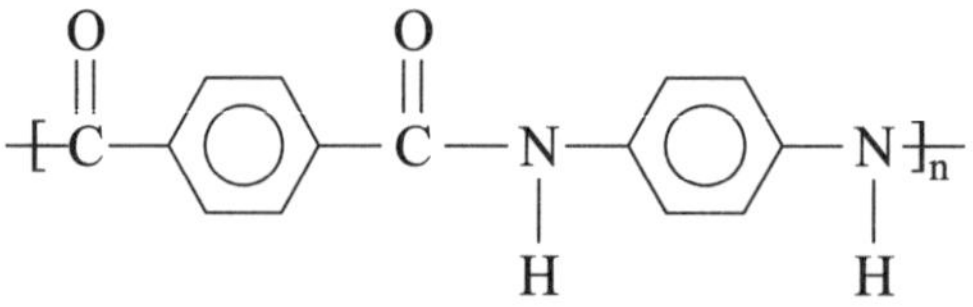

```
   O              O
   ||             ||
 -[C--<ring>--C---N--<ring>--N-]n-
                  |          |
                  H          H
```

Kevlar is a very crystalline polymer. Its dissolution in any solvent is very difficult and it starts melting above 500 °C.

Nomex, on the other hand, has meta-phenylene groups, i.e., the amide groups are attached to the phenyl ring at the 1 and 3 positions.

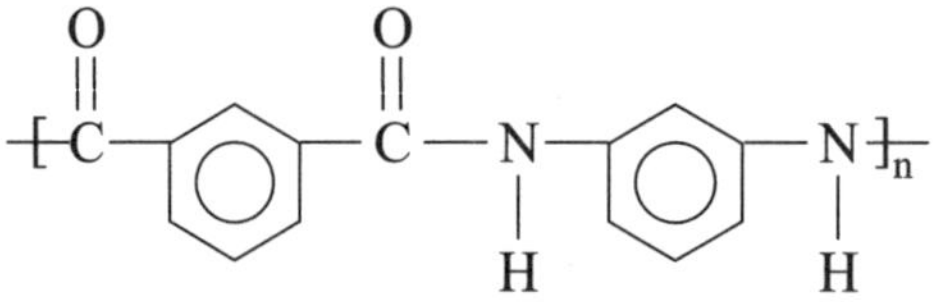

```
   O              O
   ||             ||
 -[C--<ring>--C---N--<ring>--N-]n-
                  |          |
                  H          H
```

Polyesters

Polyesters have hydrocarbon backbones which contain ester linkages, hence the name.

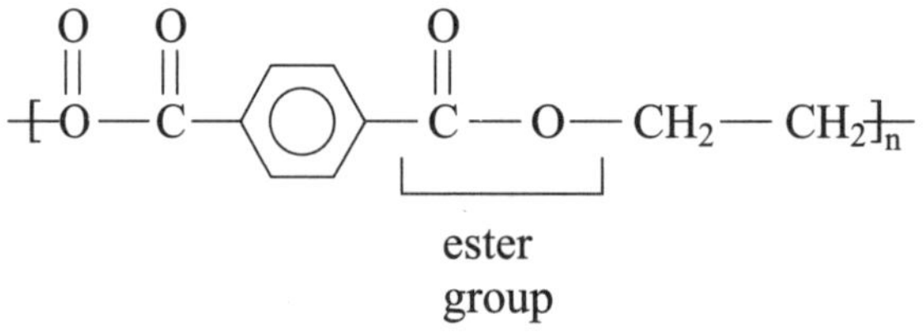

```
       O              O
       ||             ||
 -[O---C--<ring>--C---O---CH2---CH2-]n-
                 |_________|
                    ester
                    group
```

Several different materials can be used to make polyester:

1. Diacid and a diol:

$$HO-\overset{O}{\overset{\|}{C}}-R_1-\overset{O}{\overset{\|}{C}}-OH+HO-R_2-OH \rightarrow -[O-\overset{O}{\overset{\|}{C}}-R_1-\overset{O}{\overset{\|}{C}}-O-R_2-]-+H_2O$$

2. Diester and a diol:

$$RO-\overset{\overset{O}{\|}}{C}-R_1-\overset{\overset{O}{\|}}{C}-OR+HO-R_2-OH\rightarrow -[-O-\overset{\overset{O}{\|}}{C}-R_1-\overset{\overset{O}{\|}}{C}-O-R_2-]-+ROH$$

3. Transesterification of glycol ester and a diacid:

$$O-\overset{\overset{O}{\|}}{C}-R_1-\overset{\overset{O}{\|}}{C}-Cl+R_3-O-\overset{\overset{O}{\|}}{C}-R_2-\overset{\overset{O}{\|}}{C}-O-R_3 \rightarrow$$

$$-[-O-\overset{\overset{O}{\|}}{C}-R,-\overset{\overset{O}{\|}}{C}-O-R_2-]-+R_3-\underset{\underset{O}{\|}}{C}-O-H$$

4. Diacid chloride and a diol:

$$O-\overset{\overset{O}{\|}}{C}-R_1-\overset{\overset{O}{\|}}{C}-O+HO-R_2-OH \rightarrow$$

$$-[-O-\overset{\overset{O}{\|}}{C}-R_1-\overset{\overset{O}{\|}}{C}-O-R_2-]-+HO$$

5. Transesterification of ethylene carbonate and a diacid:

$$\begin{matrix} & O \\ & /\!/ \\ C- & C \\ & \backslash \\ | & O \\ & / \\ C- & C \\ & \backslash\!\backslash \\ & O \end{matrix} + HO-\overset{\overset{O}{\|}}{C}-R-\overset{\overset{O}{\|}}{C}-OH \rightarrow$$

$$-[-O-\overset{\overset{O}{\|}}{C}-R-\overset{\overset{O}{\|}}{C}-O-C-C-]-+H-\overset{\overset{O}{\|}}{C}-OH$$

6. Self-condensation of hydroxy acid:

$$HO-R_1-\overset{\overset{O}{\|}}{C}-OH\rightarrow -[-O-R_1-\overset{\overset{O}{\|}}{C}-O-]-$$

7. Condensation of anhydride and diol:

$$\text{C}_6\text{H}_4(\text{CO})_2\text{O} + \text{HO}-\text{CH}_2-\text{CH}_2-\text{OH} \longrightarrow \left[-\text{O}-\overset{\text{O}}{\overset{\|}{\text{C}}}-\text{C}_6\text{H}_4-\overset{\text{O}}{\overset{\|}{\text{C}}}-\text{O}-\text{CH}_2-\text{CH}_2- \right]$$

8. Condensation of diphenol and diacid:

$$\text{HO}-\text{C}_6\text{H}_4-\text{O}-\text{C}_6\text{H}_4-\text{OH} + \text{HO}-\overset{\text{O}}{\overset{\|}{\text{C}}}-\text{CH}_2-\text{CH}_2-\text{CH}_2-\text{CH}_2-\overset{\text{O}}{\overset{\|}{\text{C}}}-\text{OH} \longrightarrow \left[-\text{C}_6\text{H}_4-\text{O}-\text{C}_6\text{H}_4-\text{O}-\overset{\text{O}}{\overset{\|}{\text{C}}}-\text{CH}_2-\text{CH}_2-\text{CH}_2-\text{CH}_2-\overset{\text{O}}{\overset{\|}{\text{C}}}-\text{O}- \right]$$

9. Condensation of diacetate of diphenol and diacid:

$$\overset{+}{\text{M}}\,\overset{-}{\text{O}}-\overset{\text{O}}{\overset{\|}{\text{C}}}-\text{O}-\text{C}_6\text{H}_4-\text{O}-\text{C}_6\text{H}_4-\text{O}-\overset{\text{O}}{\overset{\|}{\text{C}}}-\text{O}^-\,\overset{+}{\text{M}} \longrightarrow +\text{HO}-\overset{\text{O}}{\overset{\|}{\text{C}}}-\text{CH}_2-\text{CH}_2-\text{CH}_2-\text{CH}_2-\overset{\text{O}}{\overset{\|}{\text{C}}}-\text{OH}$$

$$\left[\!\!-\!\bigcirc\!\!-\!O\!-\!\bigcirc\!\!-\!O-\overset{O}{\overset{\|}{C}}-\underset{H}{\overset{H}{\underset{|}{\overset{|}{C}}}}-\underset{H}{\overset{H}{\underset{|}{\overset{|}{C}}}}-\underset{H}{\overset{H}{\underset{|}{\overset{|}{C}}}}-\underset{H}{\overset{H}{\underset{|}{\overset{|}{C}}}}-\overset{O}{\overset{\|}{C}}-O\!-\!\right]$$

10. Condensation of disodium salt of diphenol and diacid chloride:

$$\overset{+}{M}\overset{-}{O}\!-\!\bigcirc\!-\!O\!-\!\bigcirc\!-\!O^{-}\overset{+}{M} + HO-\overset{O}{\overset{\|}{C}}-\underset{H}{\overset{H}{\underset{|}{\overset{|}{C}}}}-\underset{H}{\overset{H}{\underset{|}{\overset{|}{C}}}}-\underset{H}{\overset{H}{\underset{|}{\overset{|}{C}}}}-\underset{H}{\overset{H}{\underset{|}{\overset{|}{C}}}}-\overset{O}{\overset{\|}{C}}-OH$$

$$\downarrow$$

$$\left[\!\!-\!\bigcirc\!\!-\!O\!-\!\bigcirc\!\!-\!O-\overset{O}{\overset{\|}{C}}-\underset{H}{\overset{H}{\underset{|}{\overset{|}{C}}}}-\underset{H}{\overset{H}{\underset{|}{\overset{|}{C}}}}-\underset{H}{\overset{H}{\underset{|}{\overset{|}{C}}}}-\underset{H}{\overset{H}{\underset{|}{\overset{|}{C}}}}-\overset{O}{\overset{\|}{C}}-O\!-\!\right]$$

Some important examples of industrial polyesters are terelene, kodel etc.

Saturated Polyesters

Terelene

It is polyethylene terephthalate (PET) and is also known as dacron. It is obtained by the condensation polymeriaztion of terephthalic acid or its ester and ethylene glycol under vacuum at high temperatures.

$$HOCH_2\text{–}CH_2OH \quad + \quad ROOC\!-\!\bigcirc\!-\!COOR$$

Ethylene glycol

R = H Terepthalic acid
R = C_2H_5 Diethyltere phthalic

$$\downarrow$$

$$\left[\!-\overset{O}{\overset{\|}{C}}-\bigcirc-\overset{O}{\overset{\|}{C}}-O-CH_2-CH_2-O-\right]_n$$

Terelene

The ester groups in the polyester chain are polar, with the carbonyl oxygen atom having a somewhat negative charge and the carbonyl carbon atom having a somewhat positive charge. The positive and negative charges of different ester groups are attracted to each other. This allows the ester groups of nearby chains to line up with each other in crystal form, which is why they can form strong fibers. Due to low glass transition temperature of PET it cannot be used for making bottles and jars which need to be sterilized at high temperature.

Polyethylenenaphthalate (PEN)

There is a new kind of polyester that has high glass transition temperature, just the thing needed for jelly jars and returnable bottles. It is poly(ethylene naphthalate), or PEN.

$$\left[O - \overset{O}{\overset{\|}{C}} - C_{10}H_6 - \overset{O}{\overset{\|}{C}} - O - CH_2 - CH_2 \right]_n$$

PEN has a higher glass transition temperature than PET. The glass transition temperature of PEN is high enough so that it can withstand the heat of sterilizing bottle washing.

PBT and PTT

$$\left[\overset{O}{\overset{\|}{C}} - C_6H_4 - \overset{O}{\overset{\|}{C}} - O - CH_2 - CH_2 - CH_2 - CH_2 - O \right]_n$$

Poly(butylene terephthalate

$$\left[\overset{O}{\overset{\|}{C}} - C_6H_4 - \overset{O}{\overset{\|}{C}} - O - CH_2 - CH_2 - CH_2 - O \right]_n$$

Poly (trimethylene terephthalate)

They are usually used for the same type of things as PET, but in some cases these perform better.

Kodel

It is a copolymer of terephthalic acid and cyclohexane dimethanol.

$$HOH_2C-C_6H_{10}-CH_2OH + HOOC-C_6H_4-COOH$$

Cyclohexane dimethomit　　　　Terephtlalic acid

$$\downarrow$$

$$-O-H_2C-C_6H_{10}-CH_2-O-\overset{\overset{O}{\|}}{C}-C_6H_4-CH_2-O-$$

Kodel

It is used for making dress material. It has good crease resistance qualities and retention of shape and size during laundering.

Unsaturated Polyesters

Unsaturated polyesters are obtained from phthalic acid, maloic acid, ethylene glycol and butanadial. The double bond present in the polyester is used for cross-linking with styrene using free radical catalysts.

The unsaturated polyster-styrene crossed linked polymers are used as resin matrix in fiber reinforced plastic structures.

Aliphatic unsaturated polyesters find the same industrial application. Unsaturation is introduced in the resin molecule by using an unsaturated dicarboxylic acid such as malaic acid. Thus the conduction of ethylene glycol with maloic acid give a polyester of the type

$$nHOOC-CH=CH-COOH \quad +- \quad nHO-CH_2-CH_2-OH \longrightarrow$$

Maloic acid　　　　Ethylene glycol

$$H-\!\!\left[O-\underset{\underset{O}{\|}}{C}-CH=CH-\underset{\underset{O}{\|}}{C}-O-CH_2-CH_2\right]_n OH + H_2O$$

High molecular weight cosine of this type is obtained by continuous removal of water formed in the reaction.

Polycarbonates

Polycarbonate gets its name from the carbonate groups in its backbone chain. Polycarbonate, or specifically polycarbonate of bisphenol A, is a clear plastic used to make shatterproof windows, aircraft canopies, windscreens, lightweight eyeglass lenses, bullet-resistant glass and household items such as bottles, compact discs, and DVDs.

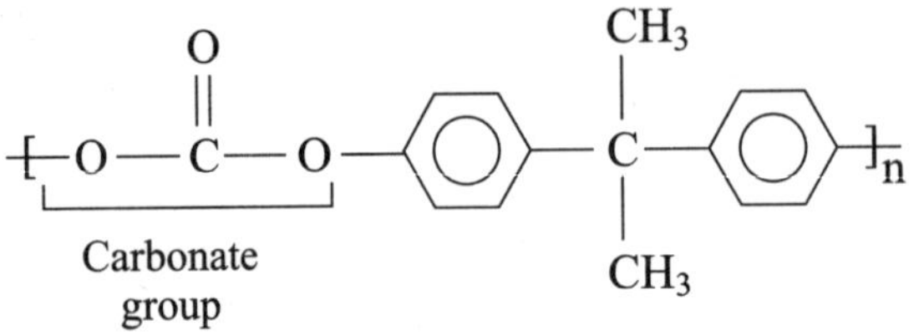

$$-\!\!\left[O - \overset{\overset{\displaystyle O}{\|}}{C} - O - \bigcirc - \underset{\underset{\displaystyle CH_3}{|}}{\overset{\overset{\displaystyle CH_3}{|}}{C}} - \bigcirc \right]_n$$

Carbonate group

Polycarbonate of bisphenol A is made by its reaction with phosgene. This starts out with the reaction of bisphenol A with sodium hydroxide to get the sodium salt of bisphenol A.

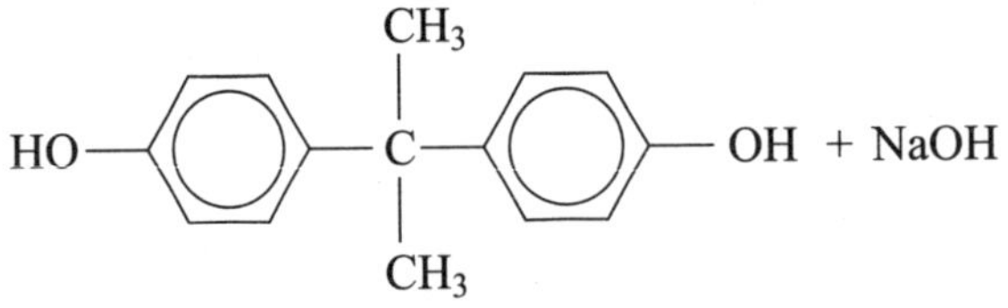

$$HO - \bigcirc - \underset{\underset{\displaystyle CH_3}{|}}{\overset{\overset{\displaystyle CH_3}{|}}{C}} - \bigcirc - OH \; + \; NaOH$$

bisphenol A

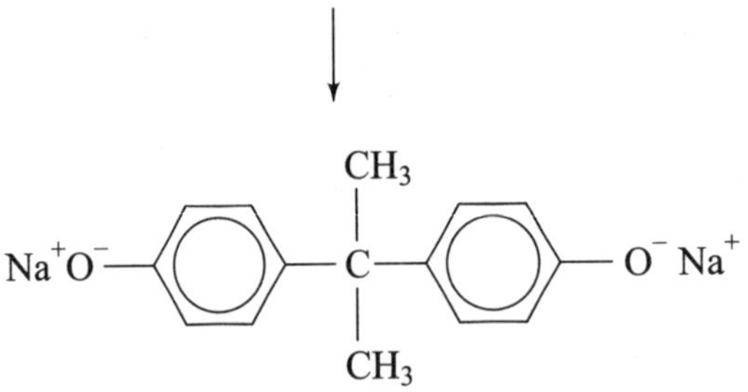

$$\downarrow$$

$$Na^{+}O^{-} - \bigcirc - \underset{\underset{\displaystyle CH_3}{|}}{\overset{\overset{\displaystyle CH_3}{|}}{C}} - \bigcirc - O^{-}Na^{+}$$

sodium salt of bisphenol A

The sodium salt of bisphenol A is then reacted with phosgene to produce the polycarbonate.

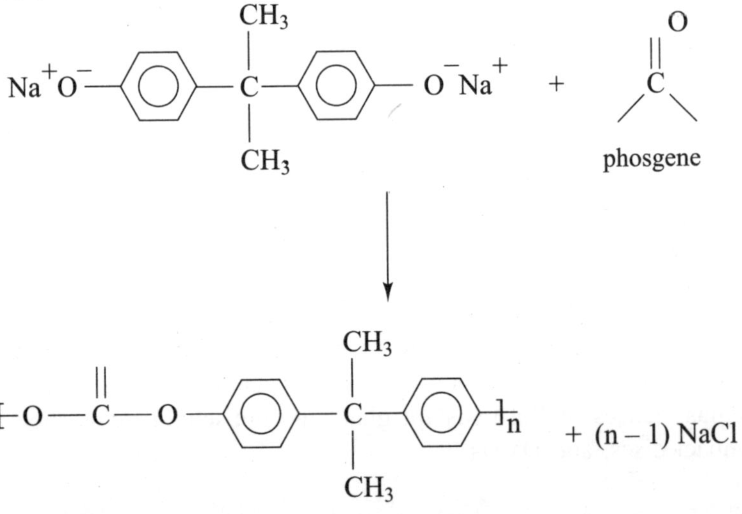

$$Na^{+}O^{-} - \bigcirc - \underset{\underset{\displaystyle CH_3}{|}}{\overset{\overset{\displaystyle CH_3}{|}}{C}} - \bigcirc - O^{-}Na^{+} \; + \; \overset{\overset{\displaystyle O}{\|}}{C}\diagup\diagdown$$

phosgene

$$\downarrow$$

$$-\!\!\left[O - \overset{\|}{C} - O - \bigcirc - \underset{\underset{\displaystyle CH_3}{|}}{\overset{\overset{\displaystyle CH_3}{|}}{C}} - \bigcirc \right]_n + (n-1)\,NaCl$$

Polycarbonate of Bisphenol A

Polypropylene carbonate

Polypropylene carbonate (PPC), also called poly(propylene carbonate), is a polycarbonate plastic material, a copolymer of carbon dioxide and propylene oxide. It is made by reacting the monomers in presence of zinc glutarate catalyst.

Polypropylene carbonate is soluble in polar solvents, e.g., acetone, methylethyl ketone, ethyl acetate, dichloromethane and chlorinated hydrocarbons. It is insoluble in ethyl alcohol, ethylene glycols, water, and aliphatic hydrocarbons. Stable emulsions in water are available commercially.

Polypropylene carbonate is used as a binder, and to increase the toughness of some epoxy resins. It is unsuitable as an oxygen barrier, and therefore is used in applications where diffusion of oxygen through the plastic is desired. Composites of polypropylene carbonate with starch (PPC/starch) can be used as a biodegradable plastic.

In ceramic industry, polypropylene carbonate and polyethylene carbonate are used as sacrificial binders, especially alumina and silicon carbide; the polymers get burned during sintering as they decompose completely above 300°C. Polypropylene carbonate binders are amorphous, with glass transition temperature of 40°C. Polyethylene carbonate has glass transition temperature of 25°C.

Polysulfone

Polysulfone or PSU is a polymer thermoplastic material. It is tough, rigid, high-strength, and transparent, retaining its properties between −100°C and +150°C. It has a very high dimensional stability.

Poly (phenyl sulfone)

Poly (phenyl sulfone) is so stiff that it does not have a glass transition temperature. It stays hard as a rock right up to its decomposition at around 500°C.

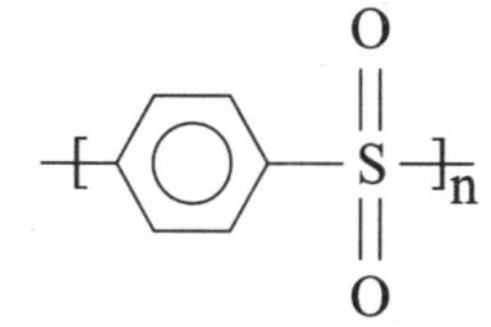

$$\left[-C_6H_4-\overset{O}{\underset{O}{\overset{\|}{\underset{\|}{S}}}}- \right]_n$$

poly(phenylene sulfone)

Polyether sulfones (PES)

In order to make polyphenyl sulfones processable, flexible groups are introduced in the backbone chain, namely ether linkages. This is how, poly(ether sulfones) are produced. They are high performance polymers and act a lot like polycarbonates, but they are a lot more heat resistant. They are used to make things like cookware and medical products that need to be sterilized between uses as they can stand up well to water and steam.

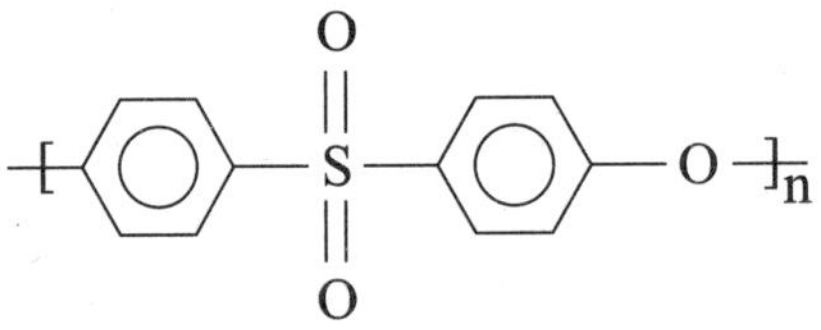

Polyether Sulfone

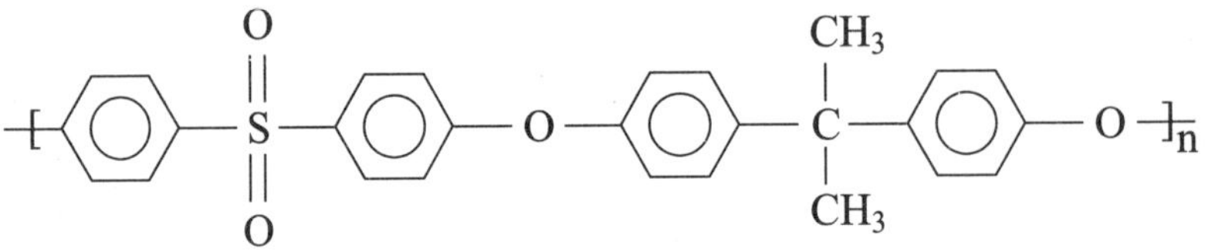

Polyetherether Sulfone

Polyketones

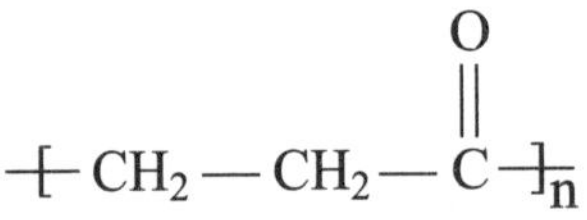

$$\left[CH_2 - CH_2 - \overset{\overset{\displaystyle O}{\|}}{C} \right]_n$$

Polyketones are very highly crystalline, that is, the polymer chains are packed together in a very orderly fashion. This makes the polyketones very strong. There production is very cost effective.

Polyetheretherketones

Polyetheretherketones (PEEK) are obtained from aromatic dihalides and bisphenolate salts by nucleophilic substitution. The bisphenolate salt is formed *in situ* from bisphenol and either added sodium or added alkali metal carbonate or hydroxide.

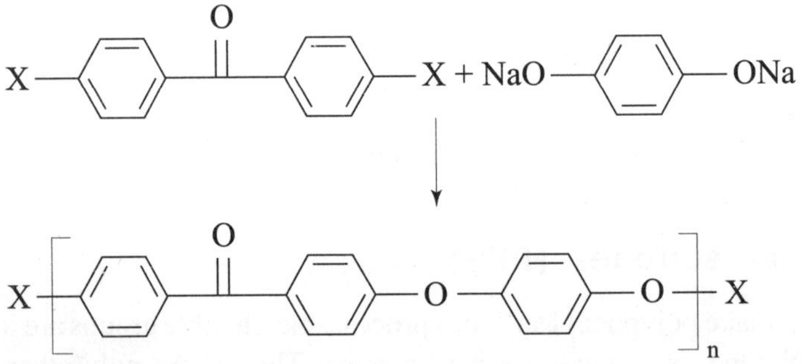

PEEK

PEEK is a thermoplastic with extraordinary mechanical properties. PEEK is partially crystalline, and has a glass transition temperature of 143 °C, a melting temperature of 334 °C and is highly resistant to thermal degradation. The material is also resistant to both organic and aqueous environments, and is used in bearings, piston parts, pumps, compressor plate valves, and cable insulation applications. PEEK is considered an advanced biomaterial used in medical implants, often in reinforced format using biocompatible fiber fillers such as carbon. Also in carbon fiber reinforced form, PEEK has come under consideration as an aerospace structural material due to its high strength-to-weight ratio. PEEK also exhibits good chemical resistance in many environments, including alkalis (i.e. sodium, potassium and ammonium hydroxides), aromatic hydrocarbons, alcohols (i.e. ethanol, propanol), greases, oils and halogenated hydrocarbons.

Polyethylene glycol

Polyethylene glycol (PEG) is the most commercially important polyethers. It refers to an oligomer or polymer of ethylene oxide with low molecular weight while polyethylene oxide is used for higher molecular weights. PEG generally is a liquid while PEO is a low-melting solid. Both are prepared by polymerization of ethylene oxide. While they find use in different applications and have different physical properties (i.e. viscosity) due to chain length effects, their chemical properties are nearly identical. The melting point is around 68 °C. PEG is soluble in water, methanol, benzene, dichloromethane and is insoluble in diethyl ether and hexane. It is coupled to hydrophobic molecules to produce non-ionic surfactants.

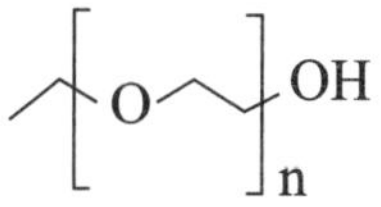

O OH n

Polyethylene glycol is non-toxic and is used in a variety of products. It is the basis of a number of laxatives and many skin creams. Polyethylene glycol with added electrolytes is used for bowel preparation and drug overdoses. PEG is used in a number of toothpastes as a dispersant; it binds water and helps keep gum uniform throughout the toothpaste. It is also used in liquid body armor and tattoos to monitor diabetes. Polyethylene glycol is also commonly used as a polar stationary phase for gas chromatography, as well as a heat transfer fluid in electronic testers. PEG is included in many or all formulations of the soft drink.

PEO (polyethylene oxide)

It is just like PEG and can serve as the separator and electrolyte solvent in lithium polymer cells. Its low diffusivity often requires high temperatures of operation, but its high viscosity even near its melting point allows very thin electrolyte layers. While crystallization of the polymer can degrade performance, many of the salts used to carry charge can also serve as a kinetic barrier to the formation of crystals. Such batteries carry greater energy for their weight than other lithium ion battery technologies.

Polypropylene glycol

Polypropylene glycol or polypropylene oxide is the polymer of propylene glycol. Chemically it is a polyether. The term polypropylene glycol or PPG is reserved for low to medium range molar mass polymer when the nature of the end-group which is usually a hydroxyl group, still matters. The oxide is used for high molar mass polymer when end groups no longer affect polymer properties.

Polypropylene glycol is produced by anionic ring-opening polymerization of propylene oxide. The initiator is an alcohol and the catalyst a base, usually potassium hydroxide. When the initiator is ethylene glycol or water the polymer is linear. With a multifunctional initiator like glycerine, pentaerythritol or sorbitol the polymer branches out.

N (propylene oxide, CH_3) —[HO–CH₂–CH(OH)–CH_3, KOH]→ HO–CH(CH_3)–CH₂–O–[CH₂–CH(CH_3)–O]$_n$–CH₂–CH(CH_3)–OH

PPG has many properties in common with polyethylene glycol. The polymer is a liquid at room temperature. Solubility in water decreases rapidly with increasing molar mass. Secondary hydroxyl groups in PPG are less reactive than primary hydroxyl groups in polyethylene glycol. PPG is used in many formulations for polyurethanes. It is used as a rheology modifier.

Polyurethane

$O{=}C{=}N-C_6H_4-CH_2-C_6H_4-N{=}C{=}O + HO-CH_2-CH_2-OH$

4,4! diisocyanatodiphenylmethane

ethylene glycol

↓

$\left[-\overset{O}{\overset{\|}{C}}-\underset{H}{\underset{|}{N}}-C_6H_4-CH_2-C_6H_4-\underset{H}{\underset{|}{N}}-\overset{O}{\overset{\|}{C}}-O-CH_2-CH_2-O- \right]_n$

A polyurethane

A polyurethane is any polymer consisting of a chain of organic units joined by urethane links. Polyurethane products are often called urethanes. They should not be confused with the specific substance urethane, also known as ethyl carbamate. Polyurethanes are not produced from ethyl carbamate, nor do they contain it.

The main polyurethane producing reaction is between a diisocyanate (aromatic and aliphatic types are available) and a polyol, typically a polyethylene glycol or polyester polyol, in the presence of catalysts. Polyurethane can be made in a variety of densities and hardness by varying the type of monomer(s) used and adding other substances to modify their characteristics, notably density, or enhance their performance.

Sometimes, the dialcohol is replaced with a diamine, to get a polyurea. It contains a urea linkage, rather than a urethane linkage. But polyureas are usually called polyurethanes.

$$O{=}C{=}N-C_6H_4-CH_2-C_6H_4-N{=}C{=}O \;+\; H_2N-CH_2-CH_2-NH_2$$

ethylene diamine

4, 4– diisocyanatodiphenylmethane

$$\downarrow$$

$$\left[C(=O)-NH-C_6H_4-CH_2-C_6H_4-NH-C(=O)-NH-CH_2-CH_2-NH \right]_n$$

A polyurea

Though the properties of the polyurethane are determined mainly by the choice of polyol, the diisocyanate exerts some influence. The cure rate is influenced by the functional group reactivity and the number of isocyanate groups—the functionality. The mechanical properties are influenced by the functionality and the molecular shape. The choice of diisocyanate also affects the stability of the polyurethane upon exposure to light. Polyurethanes made with aromatic diisocyanates yellow with exposure to light, whereas those made with aliphatic diisocyanates are stable.

Softer, elastic, and more flexible polyurethanes result when linear difunctional polyethylene glycol segments, commonly called polyether polyols, are used to create the urethane links. This strategy is used to make spandex elastomeric fibers and soft rubber parts, as well as foam rubber. More rigid products result if polyfunctional polyols are used, as these create a three-dimensional cross-linked structure which again, can be in the form of a low-density foam.

Polyurethane products have many uses. It is widely used in flexible and rigid foams, durable elastomers and high performance adhesives and sealants, fibers, seals, gaskets, condoms, carpet underlayment, and hard plastic parts. Over three quarters of the global consumption of polyurethane products is in the form of foams, with flexible and rigid types being roughly equal in market size. In both cases, the foam is usually behind other materials: flexible foams are behind upholstery fabrics in commercial and domestic furniture; rigid foams are inside the metal and plastic walls of most refrigerators and freezers, or behind paper, metals and other surface materials in the case of thermal insulation panels in the construction sector. The precursors of expanding polyurethane foam are available in many forms, for use in insulation, sound deadening, floatation, packing material, and even cast-in-place upholstery padding. Since they adhere to most surfaces and automatically fill voids, they have become quite popular in these applications.

Polyurethane materials are used in coatings and varnishes used in furniture manufacture, carpentry or woodworking. A polyurethane varnish is frequently employed as a finishing coat to protect or seal wood. Polyurethane is used as an adhesive, especially as a wood working glue. Its main advantage over more traditional wood glues is its water resistance.

Polyurethane is also used in making solid tyres. Modern roller blading and skateboarding became economical only with the introduction of tough, abrasion-resistant polyurethane parts. Polyurethane is also used in furniture manufacture for casting soft edges around table tops and panel that are stylish, very durable and prevent injury. Flexible and semi-flexible polyurethane foams are used extensively for interior components of automobiles, in seats, headrests, armrests, roof liners and instrument panels. Some surfboards are made with a solid polyurethane core.

9.1.2 Thermosets

Thermosetting polymers are crosslinked polymers. Some thermoplastic polymers (not crosslinked) behave as thermosets due to the presence of extensive secondary bonding and highly ordered structure, e.g., PAN.

Polycarbonates

Allyl diglycol carbonate is made by polymerization of diethyleneglycol bis allylcarbonate (ADC) in the presence of diisopropyl peroxydicarbonate (IPP) catalyst. The presence of the allyl groups allows the polymer to form cross-links; thus, it is a thermoset resin. It is very different from polycarbonate of bisphenol A. It is synthesized by starting with the monomer containing two allyl groups at the ends:

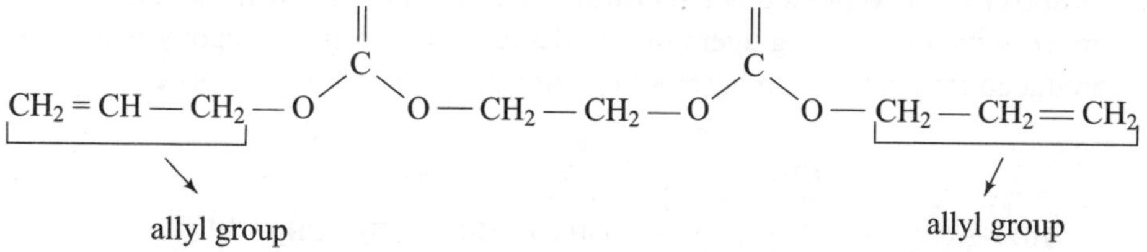

$$CH_2=CH-CH_2-O-\overset{\overset{O}{\|}}{C}-O-CH_2-CH_2-O-\overset{\overset{O}{\|}}{C}-O-CH_2-CH_2=CH_2$$

allyl group — allyl group

The presence of two allyl groups in monomer molecule ascertains that it can be polymerized free radically and the polymer will be crosslinked. This crosslinking makes the material very strong. This polymer does break nearly as easily as glass does. The crosslinked polycarbonate looks like this:

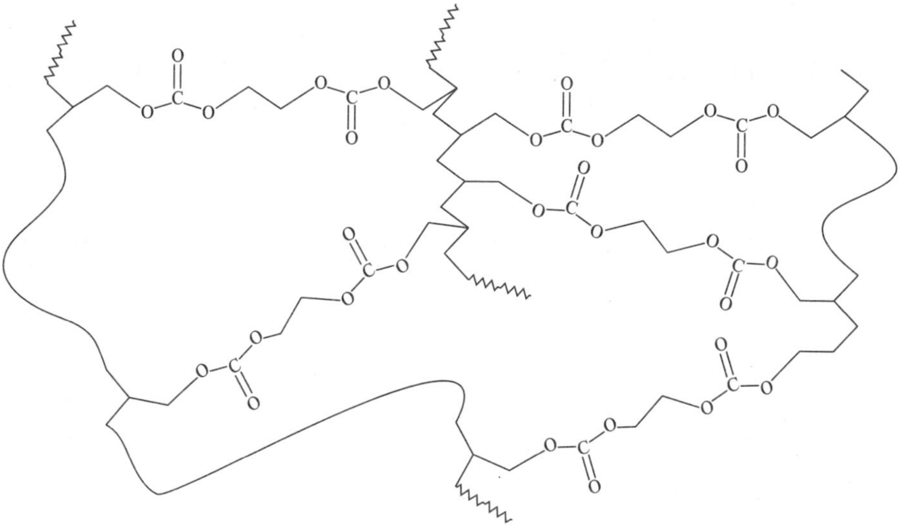

It is commonly used in the manufacture of eyeglass lenses. An alternative use—a purified version is used to measure neutron radiation, a type of ionizing radiation, in neutron dosimetry. It has high abrasion resistance, about half the weight of glass and index of refraction only slightly lower than that of the crown glass, making it an advantageous material for lenses of eyeglasses and sunglasses. A wide range of colors can be achieved by dyeing of the surface or the bulk of the material. It is also resistant to most of the solvents and other chemicals, to gamma radiation, to ageing, and to material fatigue. It can withstand the small hot sparks from welding. It can be used continuously in temperatures up to 100 °C and up to one hour in 130°C.

Epoxy Resins

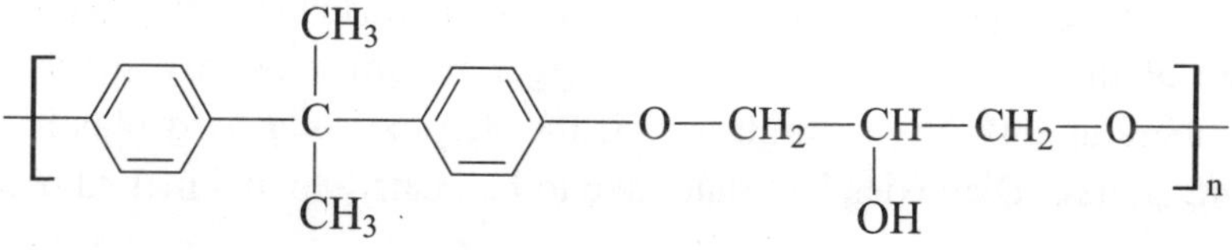

$$\left[-C_6H_4-\underset{CH_3}{\overset{CH_3}{C}}-C_6H_4-O-CH_2-\underset{OH}{CH}-CH_2-O- \right]_n$$

Epoxy or polyepoxide is a thermosetting epoxide polymer that cures when mixed with a catalyzing agent or "hardener". Most common epoxy resins are produced from a reaction between epichlorohydrin and bisphenol-A:

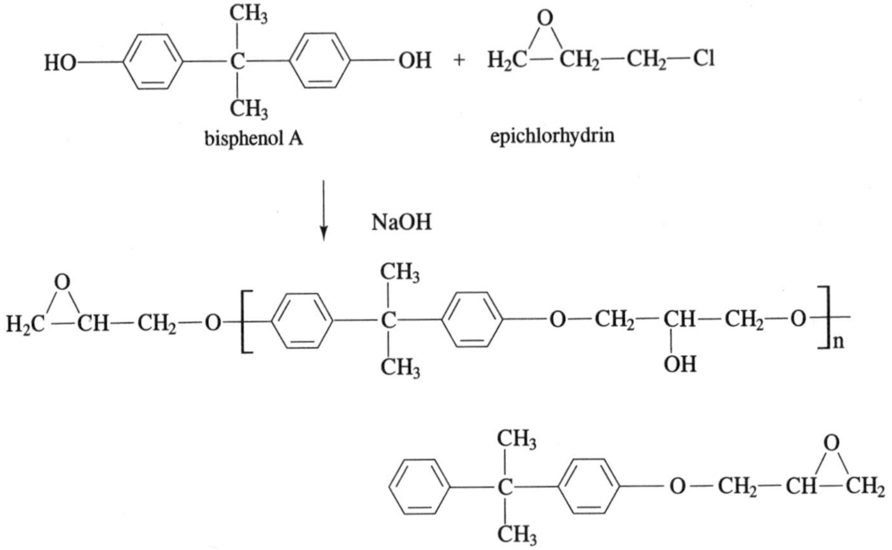

bisphenol A

epichlorhydrin

NaOH

The first part of the two-part epoxy is a low-molecular-weight polymer with epoxy groups at each end. In these prepolymers *n* can be a high as 25. The diepoxy in two-part epoxy adhesive is more likely a small molecule with two epoxy groups as shown below:

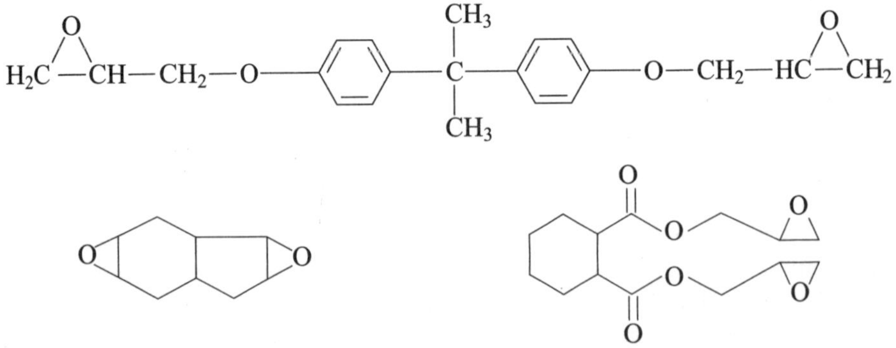

If two-part epoxy adhesive is two tubes of liquid, than the diepoxy is probably one of the small molecules. The bigger prepolymers are solids at room temperature. If *n* is as high as 25, the diepoxy is a hard plastic at room temperature. Diepoxies like that have to be heated up and melted before they can be mixed with that other part of the two-part epoxy mix. The second part is a diamine: When the two parts, the diepoxy and the diamine, are mixed, they join together, in such a way that links all the diepoxy and diamine molecules together, like this:

Diamine

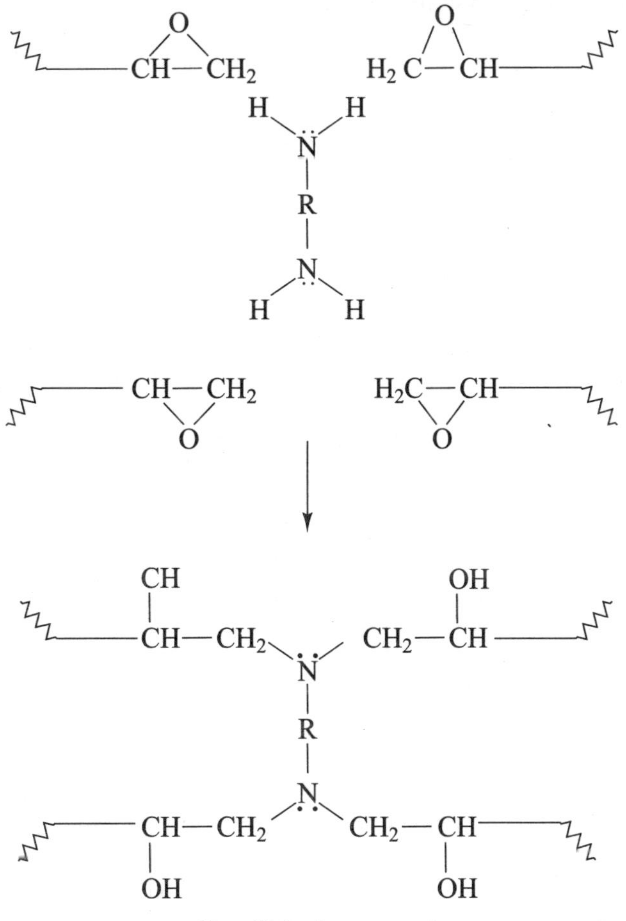

Crosslinked epoxy resin

Epoxy resins make great adhesives, and are one of the few adhesives that can be used on metals. But they are also used for things like protective coatings, and as materials in things like electronic circuit boards and for patching holes in concrete pavement.

Polyimides

A polyimide is a polymer that contains an imide group. Polyimides usually take one of two forms. The first of these is a linear structure where the atoms of the imide group are part of a linear chain. The second of these structures is a heterocyclic structure where the imide group is part of a cyclic unit in the polymer chain. Aromatic heterocyclic polyimides, like the one on the left, are typical of most commercial polyimides.

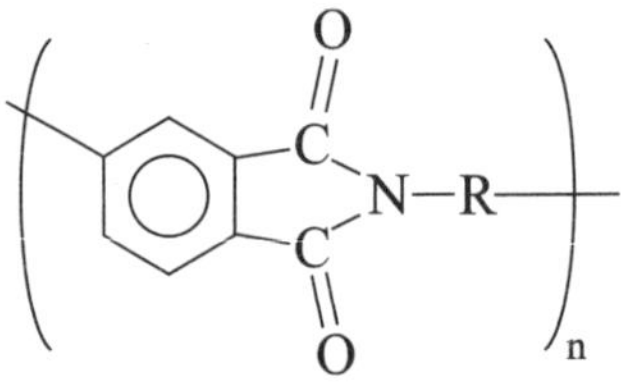

Aromatic heterocyclic polyimide 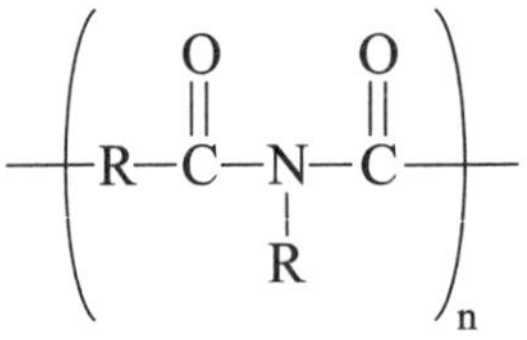   Linear polymide

Polyimides are a very interesting group of incredibly strong and astoundingly heat and chemical resistant polymers. Their strength and heat and chemical resistance are so great that these materials often replace glass and metals, such as steel, in many demanding industrial applications. Polyimides are even used in many everyday applications. Polyimides are employed for the struts and chasis in some cars as well as some parts under-the-hood because they can withstand the intense heat and corrosive lubricants, fuels, and coolants cars require. They are also used in the construction of many appliances as well as microwave cookware and food packaging because of their thermal stability, resistance to oils, greases, and fats and their transparency to microwave radiation. They can also be used in circuit boards, insulation, fibers for protective clothing, composites, and adhesives.

These properties come from strong intermolecular forces between the polymer chains. A polymer which contains a charge transfer complex consists of two different types of monomers, a donor and an acceptor. Its nitrogen groups have plenty of electrons to go around and its carbonyl groups act as the acceptors. So both the groups are held together tightly.

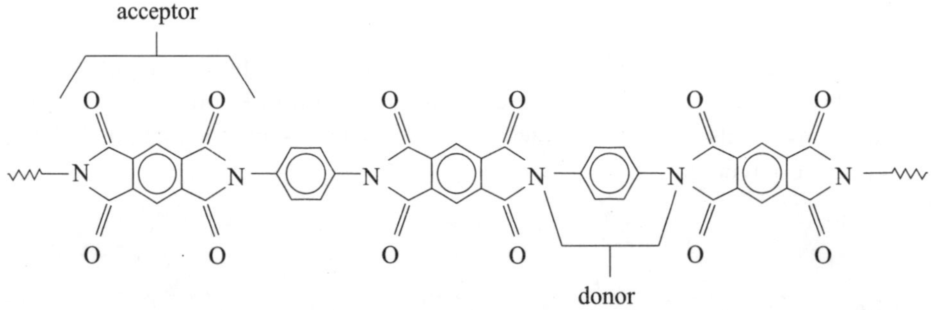

acceptor

donor

The charge transfer complex works not only between adjacent units in the polymer chain, but also between chains. The chains will stack together like strips of paper, with donors and acceptors paired up as shown below:

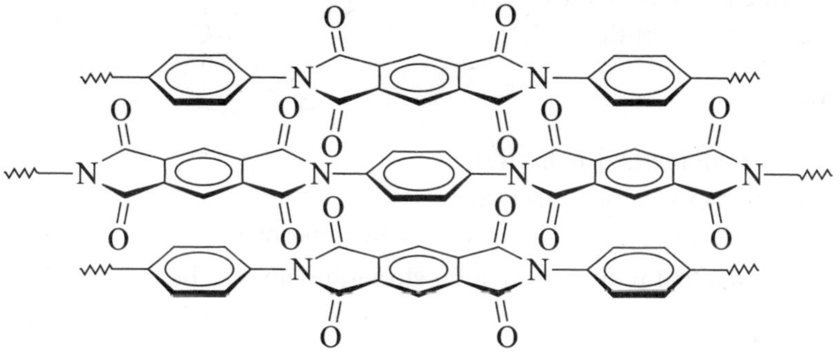

This charge transfer complex holds the chains together very tightly, not allowing them to move around very much. When things cannot move around on the molecular level, they are not capable of moving around in the whole material. This is why polyimides are so strong.

Polydicyclopentadiene

Polydicyclopentadiene is made by ring-opening metathesis polymerization (ROMP) from the monomer endo-dicyclopentadiene.

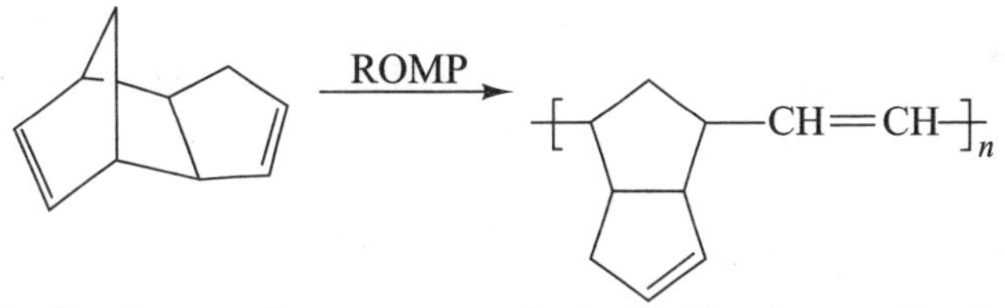

ROMP

endo-dicyclopentadiene poly (*endo*-dicyclopentadiene)

There is a double bond left in the bottom ring, which can undergo vinyl polymerization, to give a crosslinked thermoset material.

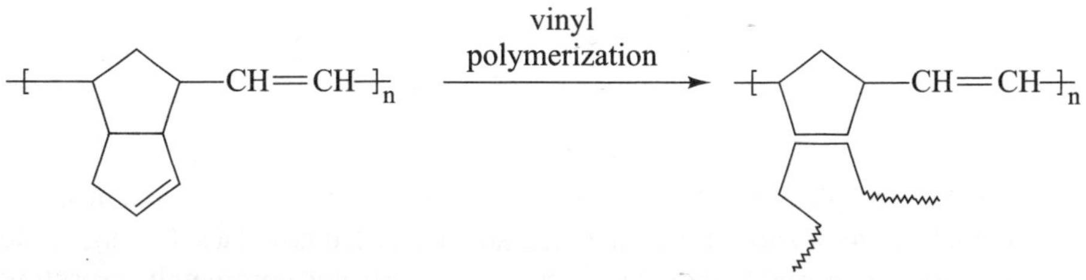

vinyl polymerization

poly (*endo*-dicyclopentadiene) Cr osslinked polymer

Polydicyclopentadiene is a polymer used to make very big things in one piece. With polydicyclopentadiene, a whole tractor cab or a whole satellite dish antenna can be made in one piece. This is because it has a very good impact resistance at low temperatures, where a lot of other polymers become brittle.

Zylon

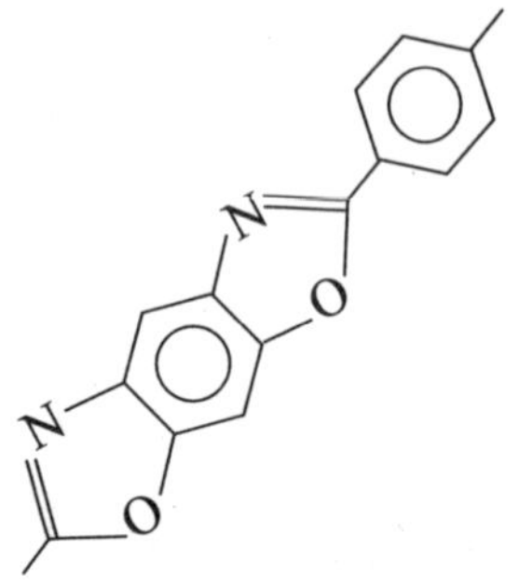

Zylon is a trademarked name for a range of thermoset polyurethane materials. Zylon is used in a number of applications that require very high strength with excellent thermal stability like Tennis racquets, Table Tennis blades and various medical applications.

Poly (p-phenylene- 2, 6- benzobisoxazole)

Phenolic Resins

They include any of various synthetic thermosetting resins, obtained by the reaction of phenols with simple aldehydes and used to make molded products, including pool and snooker balls, and as coatings and adhesives.

Phenol Formaldehyde Resin

The earliest commercial synthetic resin is based on a phenol formaldehyde resin with the commercial name Bakelite, and is formed from the elimination reaction of phenol with formaldehyde. Phenol is reactive towards formaldehyde at the *ortho* and *para* sites (sites 2, 4 and 6) allowing up to 3 units of formaldehyde to attach to the ring. This forms a hydroxymethyl phenol, which is not usually isolated in novolacs but is found in resoles. The hydroxymethyl group is capable of reacting with either another free *ortho* or *para* site, or with another hydroxymethyl group. The first reaction forms a methylene bridge, and the second forms an ether bridge.

Phenol formaldehyde resins, as a group, are formed by a step-growth polymerization reaction which may be either acid or base catalysed. The pathway the reaction follows varies depending on the catalyst type used.

Acid catalysed

Acid catalysed phenol formaldehyde resins are made with a molar ratio of formaldehyde to phenol of less than one and are called novolacs. Owing to the molar ratio of formaldehyde to phenol, they will not completely crosslink (polymerize) without the addition of a crosslinking agent. Novolacs are commonly used as photoresists.

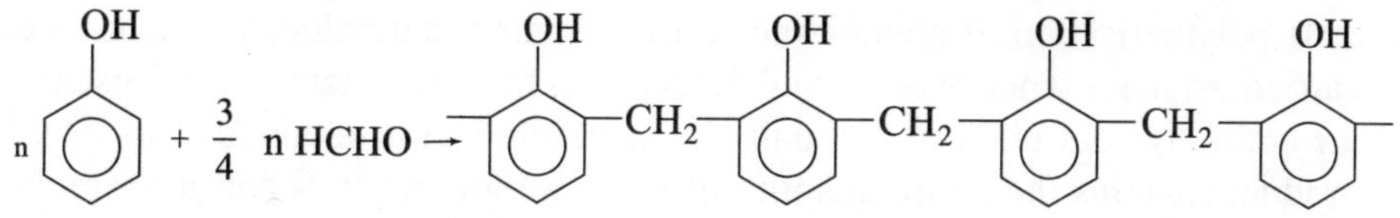

$$n\ \text{(phenol)} + \frac{3}{4}\ n\ HCHO \rightarrow$$

In case excess of formaldehyde is used and the reaction is allowed to continue (in the presence of acid catalyst), *n* cross linked three dimensional polymer called Bakelite is obtained.

Bakelite is a brand named material based on the thermosetting phenol formaldehyde resin, polyoxybenzylmethyleneglycolanhydride, formed by the reaction under heat and pressure of phenol and formaldehyde. It was used for its nonconductive and heat-resistant properties in radio and telephone casings and electrical insulators, and was also used in diverse products such as kitchenware, jewellery, pipe stems, and toys.

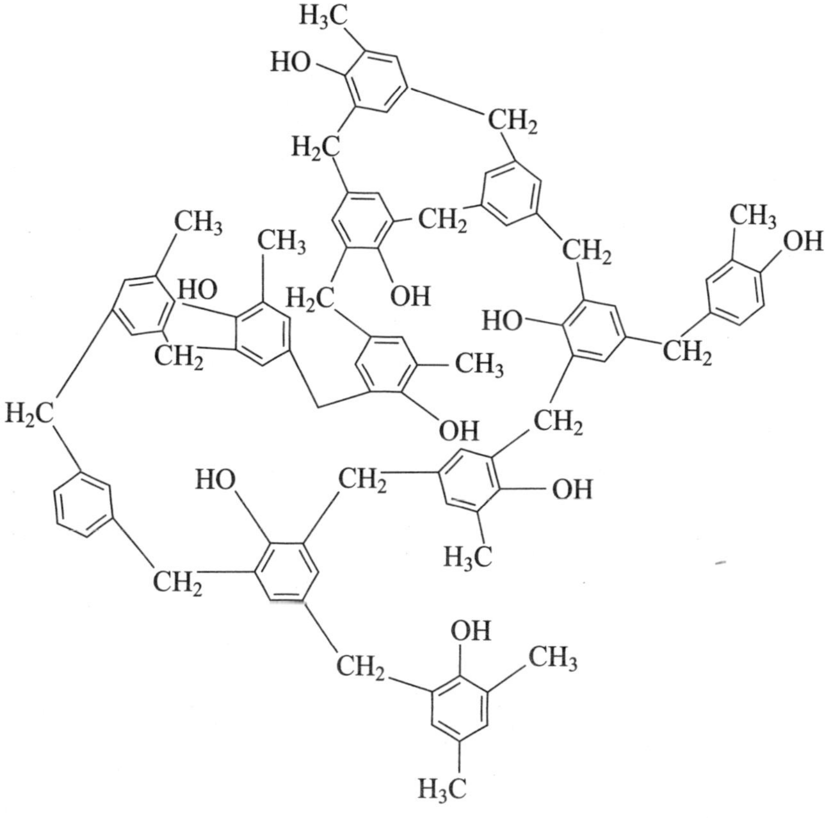

Base catalysed

Base catalysed phenol formaldehyde resins are made with a formaldehyde to phenol ratio of greater than one (usually around 1.5). Phenol, formaldehyde, water and catalyst are mixed in the desired amount, depending on the resin to be formed, and are then heated. The first part of the reaction, at around 70 °C, forms hydroxymethyl phenols. This results in a thick reddish-brown resin.

If polymerisation of phenol and formaldehyde (equimolocular snounts) is carried out in presence of alkali and the coaction rate and extent of condensation are carefully controlled so that only the linear product is obtained. This linear product contains the entire quantity of methyiol groups at *o*- and *p*-positions and on storage slow reaction continues and a hardened rusin called `Rosol' is obtained. Rosoal can, however be converted into bakelite on further reaction.

Urea-formaldehyde Resins

They are transparent thermosetting resins made from urea and formaldehyde heated in the presence of a mild base such as ammonia or pyridine. These resins are superior to phenol-formaldehyde resins. These resins are used in adhesives, finishes, and molded objects. Urea-formaldehyde resin has a high tensile strength, flexural modulus and HDT, low water absorption and mold shrinkage and higher surface hardness, elongation at break, and volume resistance.

$$O-C(NH_2)_2 + HCHO \longrightarrow O=C(NHCH_2CH)(NH_2) \xrightarrow{HCHO} O=C(NHCH_2OH)(NHCH_2OH)$$

urea Formaldehyde Monomethylol derivative Olmethylol derivative

$$O=C(NHCH_2OH)(NHCH_2OH) \quad O=C(NH_2)(NH_2) \quad O=C(NHCH_2OH)(NHCH_2NH-C(=O)NH_2)$$

Dimethylol derivative Urea

The reaction continues and finally the resin is obtained.

$$NHCHO + {}_nNH_2CONH_2 \xrightarrow[-H_2O]{} N\left[NHCO-NH-CH_2\right]_n OH$$

Urea format dehyde resin

If excess of formaldehyde is used, the H atom of the imide group in the linear polymer can be replaced by methylol group.

$$H\left[NH\,CONH-CH_2\right]_n OH \xrightarrow{CH_2O} n\left[NH\,CO\,\underset{\displaystyle CH_2OH}{\underset{|}{N}}-CH_2\right]_n OH$$

The final composition of the polymer (as in the case of phenol formaldehyde resin) depends on the ratio of urea and formaldehyde. Thus if the linear polymer is heated with excess formaldehyde, a three dimensional network of the polymer is obtained

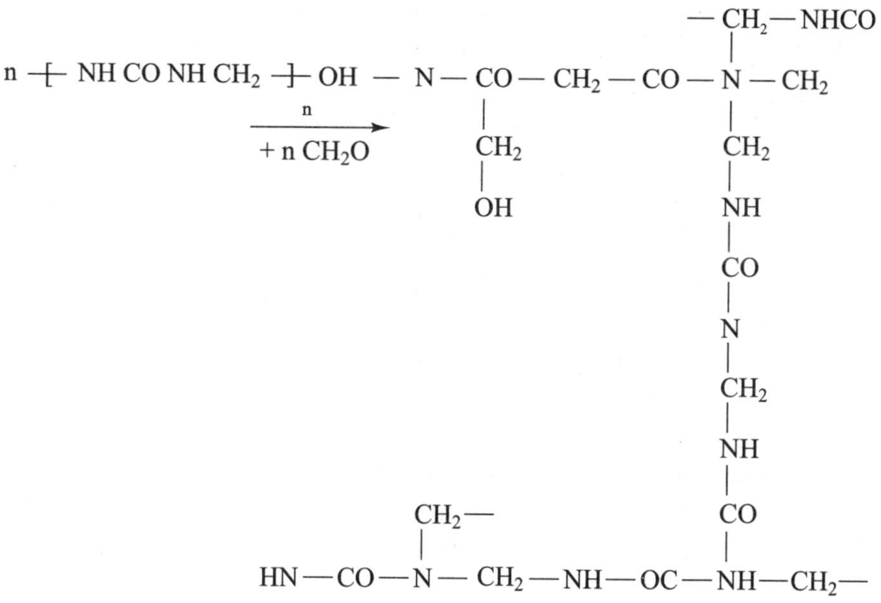

$$n\ \text{+ NH CO NH CH}_2\text{ + OH}\ \xrightarrow[+\,n\,CH_2O]{n}\ -N-CO-CH_2-CO-N(-CH_2-NHCO)-CH_2$$

$-N-$ bears CH_2-OH; the second N bears $CH_2-NH-CO-N-CH_2-NH-CO-NH-CH_2-$

$$HN-CO-N(CH_2-)-CH_2-NH-OC-NH-CH_2-$$

They are colorless and are not affected on exposure to light and are used in the manufacture of kitchenwares, laminates and as adhesives.

Melamine-formaldehyde Resins

They are produced by the reaction of melamine with formaldehyde. Melamine is produced from urea, mainly by either two methods: catalyzed gas-phase production or high pressure liquid-phase production.

The reaction of melamine with formaldehyde gives the methylol derivative of melamine.

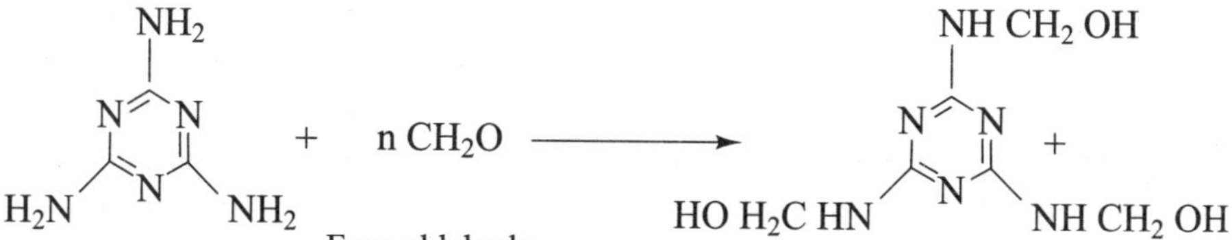

$$C_3N_3(NH_2)_3 + n\,CH_2O \longrightarrow C_3N_3(NH\,CH_2\,OH)_3 +$$

Melamine (triazine ring with NH_2, H_2N, NH_2) + Formaldehyde → triazine ring with $NH\ CH_2\ OH$, $HO\ H_2C\ HN$, $NH\ CH_2\ OH$

Formaldehyde

Melamine

Polycondensation of methylol derivative of melamine with excess of formaldehyde gives a cross linked resin.

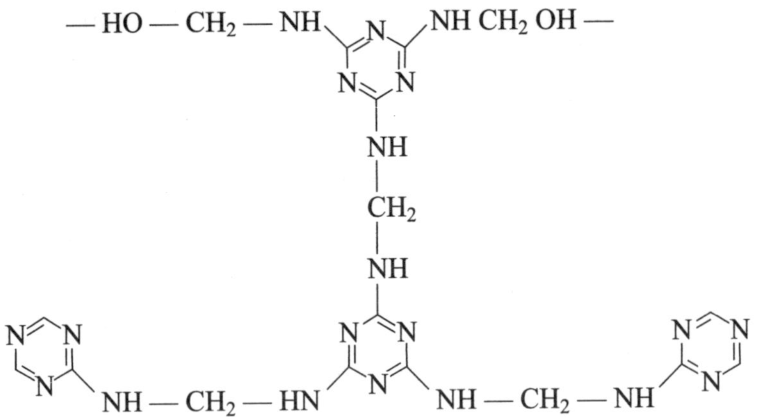

— HO — CH_2 — NH — (triazine ring: N, N, N) — NH CH_2 OH —

NH

CH_2

NH

(triazine ring: N, N, N) NH — CH_2 — HN — (triazine ring: N, N, N) — NH — CH_2 — NH — (triazine ring: N, N, N)

Melamine-formaldehyde resin is resistant to heat and chemicals and is better than urea-formaldehyde resin in hardness and tensile strength.

These are very durable thermosets. This plastic is often used in kitchen utensils or plates. Melamine tile wall panels are known as whiteboards. Melamine is also used to make decorative wall panels and is often used as a laminate. Melamine foam has an interlinking bubble format which produces a structure more like a block of microscopic fiberglass than normal foam. It is used for soundproofing, as a fire-retardant material (but not as insulation) and also as a very good cleaning product.

9.2 RUBBER POLYMERS

Polyisoprene

One of the most well known natural polymers is polyisoprene, or natural rubber. It is an elastomer as it recovers its shape after being stretched or deformed. Normally, the natural rubber is treated to give it crosslinks, which makes it an even better elastomer.

Polyisoprene is diene polymer, which is a polymer made from a monomer containing two carbon-carbon double bonds. Like most diene polymers, it has a carbon-carbon double bond in its backbone chain. Polyisoprene can be harvested from the sap of the *Hevea* tree, but it can also be made by Ziegler-Natta polymerization. This is a rare example of a natural polymer that we can make Isoprene as well as nature does.

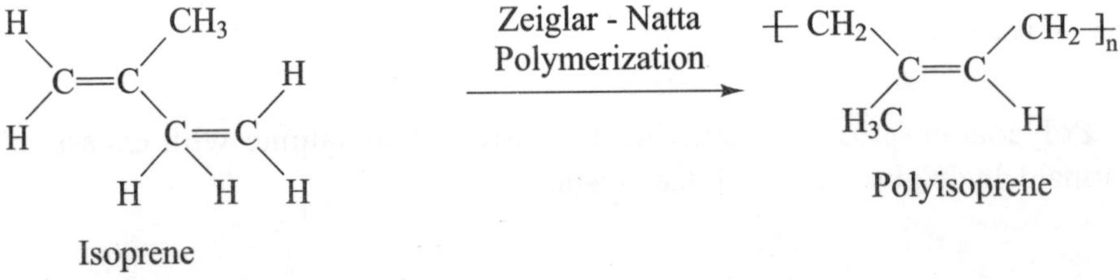

$$\underset{\text{Isoprene}}{(H)(H)C{=}C(CH_3){-}C(H){=}C(H)(H)} \xrightarrow[\text{Polymerization}]{\text{Zeiglar - Natta}} \underset{\text{Polyisoprene}}{\left[CH_2{-}C(H_3C){=}C(H){-}CH_2 \right]_n}$$

Polyisobutylene

Polyisobutylene is a synthetic rubber, or elastomer. Polyisobutylene, sometimes called butyl rubber, and other times PIB, is a vinyl polymer. It is very similar to polyethylene and polypropylene in structure, except that every other carbon is substituted with two methyl groups. It is made from the monomer isobutylene, by cationic vinyl polymerization.

$$\underset{\text{Isobutylene}}{H_2C{=}C(CH_3)_2} \xrightarrow{\text{Cationic vinyl polymerization}} \underset{\text{Polyisobutylene}}{-\!\!\left[CH_2{-}C(CH_3)_2 \right]\!\!-_n}$$

The polymerization is best carried out at temperatures from –40 °C to room temperature. This is a cationic polymerization process. There is a cure problem, though. For curing, some diene monomer such as isoprene (1 – 3%) is added. Crosslinking can be done with sulfur.

It is a rubbery polymer. Application is in adhesive and tyre industry. It is special because it is the only rubber that is gas impermeable, in other words, it is the only rubber that can hold air (balloons) for long periods of time. You may have noticed that inflated balloons go flat after a few days. This is because they are made of polyisoprene, which is not gas impermeable. Because polyisobutylene will hold air, it is used to make things like the inner liner of tires, and the inner liners of basketballs.

Polybutadiene

Polybutadiene is a diene polymer, that is, a polymer made from a monomer that contains two carbon-carbon double bonds, specifically butadiene. It is made by Ziegler-Natta polymerization.

$$CH_2{=}CH{-}CH{=}CH_2 \xrightarrow{\text{Zieglar-Natta polymerization}} -\!\!\left[CH_2{-}CH{=}CH{-}CH_2 \right]\!\!-_n$$

Polybutadiene was one of the first types of synthetic elastomer, or rubber, to be invented. It is very similar to natural rubber, polyisoprene. It is good for uses which require exposure to low temperatures. Tire treads are often made of polybutadiene copolymers. Belts, hoses, gaskets and other automobile parts are made from polybutadiene, because it stands up to cold temperatures better than other elastomers.

Polychloroprene

$$-\!\!\left[CH_2{-}C(Cl){=}C(H){-}CH_2 \right]\!\!-_n$$

Polychloroprene is usually sold under the trade name Neoprene. Chloroprene is a diene. Polychloroprene is made from the monomer chloroprene.

$$H_2C{=}C(Cl){-}CH{=}CH_2 \longrightarrow {-}\!\!\left[CH_2{-}C(Cl){=}CH{-}CH_2\right]\!_n{-}$$

Chloroprene Polychloroprene

Polychloroprene has properties similar to those of other diene polymers like polyisoprene and polybutadiene. It is especially resistant to oil. It was the first synthetic elastomer, or rubber to be a hit commercially.

Poly(styrene-butadiene-styrene) or SBS

$$-\!\left[CH_2{-}CH(C_6H_5)\right]_n\!\!-\!\left[CH_2{-}CH{=}CH{-}CH_2\right]_n\!\!-\!\left[CH_2{-}CH(C_6H_5)\right]_n-$$

Poly(styrene-butadiene-styrene), or SBS, is a type of copolymer called a block copolymer. SBS is made with some really clever chemistry called living anionic polymerization. Its backbone chain is made up of three segments. The first is a long chain of polystyrene, the middle is a long chain of polybutadiene, and the last segment is another long section of polystyrene as shown below.

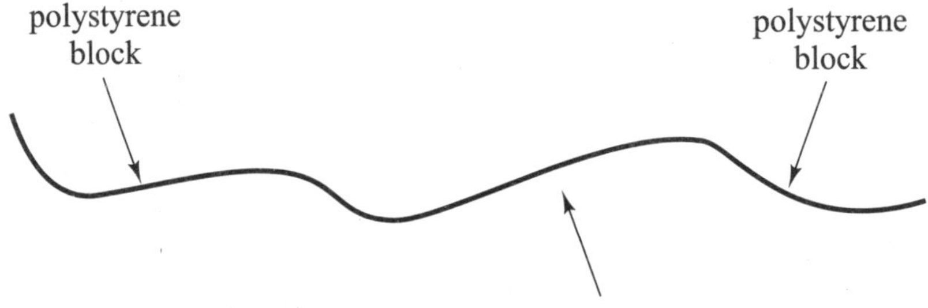

Polystyrene is a tough hard plastic, and this gives SBS its durability. Polybutadiene is rubbery, and this gives SBS its rubber-like properties. In addition, the polystyrene chains tend to clump together. When one styrene group of one SBS molecule joins one clump, and the other polystyrene chain of the same SBS molecule joins another clump, the different clumps become tied together with rubbery polybutadiene chains. This gives the material the ability to retain its shape after being stretched.

SBS is also a type of unusual material called a thermoplastic elastomer. These are materials that behave like elastomeric rubbers at room temperature, but when heated, can be processed like plastics. Most types of rubber are difficult to process because they are crosslinked. But SBS and other thermoplastic elastomers manage to be rubbery without being crosslinked, making them easy to process into nifty useful shapes. It is a hard rubber that is used for things like the soles of shoes, tire treads, and other places where durability is important.

9.3 CONDUCTING POLYMERS

There has been a considerable effort to synthesize organic polymers that arc electrically conducting. Conducting polymers offer two potential advantages over the traditional inorganic materials used as conductors. First, processing of conducting polymers by molding and other plastic processing techniques into various electrical and electronic devices and forms is easy as compared to the metallurgical processes used for inorganic conducting materials. Second, the lightweight property of polymeric materials would make certain types of applications more practical and economical. The biggest and most immediate potential application for conducting polymers is lightweight rechargeable batteries for portable devices (e.g., tools) and vehicles. Conducting polymers would serve both current carrying and ion-conduction functions by replacing traditional electrode and electrolyte substances. The other application area for conducting polymers is their use to build circuitry elements, both passive (conducting circuits) and active (p-n and Schottky junctions).

The structural requirement for a conducting polymer is a conjugated π-electron system. The typical polymers studied are based on a π-conjugated electron system constituting the polymer backbone but conducting polymers are also possible when the π-conjugated electron system is present in side chain. Polymers with conjugated π-electron systems display unusual electronic properties including high electron affinities and low ionization potentials. Such polymers are easily oxidized or reduced by charge-transfer agents (dopants) that act as electron acceptors or electron donors, respectively. This results in poly(radical-cation)- or poly(radical-anion)-doped conducting materials (p-type and n-type, respectively) with near-metallic conductivity in many cases. Electrons are removed from or added to the conjugated π-electron system to form extra holes or electrons, respectively, which carry current by wandering through the polymer chain. The dopant plays an additional role other than bringing about oxidation or reduction. It acts as the bridge or connection for carrying current between different polymer chains. Oxidative dopants include AsF_5, I_2, $AlCl_3$, and $MoCl_5$. Sodium, potassium, and lithium naphthalides are reductive dopants. The dopant, necessary to impart electrical conduction, often decreases but does not eliminate the processability of a polymer.

A number of other characteristics are required for a viable polymeric conductor. Chain orientation is needed to enhance the conducting properties of a polymeric material, especially the intermolecular conduction (i.e., conduction of current from one polymer molecule to another). This is a problem with many of the polymers that are amorphous and show poor orientation. For moderately crystalline or oriented polymers, there is a possibility of achieving the required orientation by mechanical stretching. Liquid crystal polymers would be especially advantageous for electrical conduction because of the high degree of chain orientation that can be achieved. A problem encountered with some doped polymers is a lack of stability. These materials are either oxidants or reductants relative to other compounds, especially water and oxygen.

Some of the polymers investigated as heat-resistant polymers have conjugated π-electron systems and offer potential as conducting polymers. The most common conducting polymers are polyacetylene, poly(p-phenylene sulfide), polyquinoline, polyquinoxaline, polypyrrole, thiophene, aniline, furan, carbazole, and azulene. The polymers from aniline and thiophene, together with polypyrrole, are the most promising as conducting polymers.

Classification

Conducting polymers can be classified into two types:

(i) Intrinsically conducting polymers: These polymers have extensive conjugation in the backbone which is responsible for conductance. They are of two kinds:
   a) Conducting polymers having conjugated π-electrons in the backbone, e.g., polyacetylene, polyaniline etc.
   b) Doped conducting polymers, they are either p-doped (oxidative doping) or n-doped (reductive doping).

(ii) Extrinsically conducting polymers: In this type of polymers, conductivity is due to the externally added ingredients in them. They are again of two types:
   a) Conductive element filled polymers are the binders to hold the conducting element.
   b) Blended conducting polymers are prepared by blending a conventional polymer with a conducting polymer.

Synthesis

Electrochemical and oxidative polymerizations are the two ways to prepare conducting polymers.

Electrochemical Polymerization

Passage of current through a solution results in the loss of electrons, and compounds are oxidized at the anode. Electrons are gained and compounds reduced at the cathode. This process is referred to as electrochemical polymerization when a polymer is formed. The polymerization reaction and polymer properties (conductivity and mechanical strength) are dependent on such parameters as identity and concentration of electrolyte, reaction temperature, and current density. Polymerization of thiophene, pyrrole and aniline, etc., proceeds in a similar manner.

Oxidative Polymerization

Aniline as well as pyrrole, thiophene, and other aromatics are also polymerized by chemical oxidants, e.g., pyrrole is polymerized by ferrous or silver ions and aniline by persulfate. One of the earliest attempts to synthesize a heat-resistant polymer was the oxidative-coupling polymerization of benzene to poly(p-phenylene). The reaction requires the presence of both a Lewis acid catalyst and an oxidant. The most widely used combination is aluminium chloride-cupric chloride.

Properties

The biggest advantage of conducting polymers is their processibility. They are also plastics (which are organic polymers) and therefore can combine the mechanical properties (flexibility, toughness, malleability elasticity, etc.,) of plastics with the high electrical conductivities of a doped conjugated polymer.

Applications

Applications are in rechargeable batteries, ion exchangers, optical filters and in many electronic devices. Electroluminescence (light emission) in organic compounds has been known since the early 1950's, but practical application was limited due to their low conductivity. In some cases, similar light emission is observed when a voltage is applied to a thin layer of a conductive organic polymer film. The increased conductivity of such polymers mean enough power can be put through the device at low voltages to generate practical amounts of light. This has led to the development of flat panel displays using OLEDs, solar panels and optical amplifiers.

Common classes of organic conductive polymers include poly(acetylene)s,poly(pyrrole)s, poly(thiophene)s, poly(aniline)s, poly(fluorene)s, polynaphthalenes, poly(*p*-phenylene sulfide), and poly(para-phenylene vinylene)s. Classically, these compounds are known as polyacetylene, polyaniline—blacks or melanins. The melanin pigment in animals is generally a mixed copolymer of polyacetylene, polypyrrole, and polyaniline. Some fungal melanins are pure polyacetylene.

Polyacetylene

Polyacetylene (PA) is an organic polymer that can be converted to a conducting polymer of the rigid-rod polymer host family. As prepared by a Ziegler-Natta catalyst, with high levels of catalyst, the polymer is a silver, non-conductive film. Oxidation with iodine provides a 10^8 increase in the conductivity.

The polymerization of acetylene results in a long chain of carbon atoms with alternating single and double bonds between them, each with one hydrogen atom which could be replaced by a functional group.

...CH—CH—CH—CH—CH...

Polyacetylene, prepared differently, is also called **acetylene black** or **polyacetylene black**.

Polypyrrole

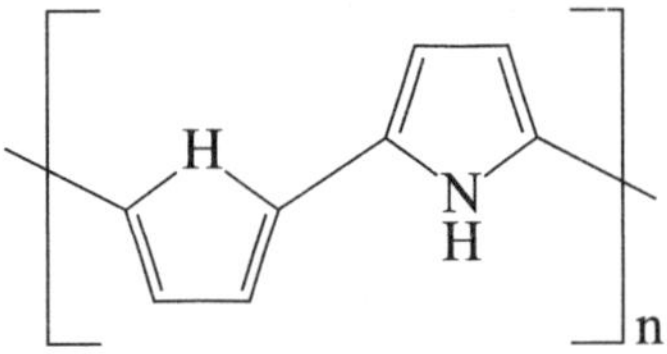

A polypyrrole (PPy) is a chemical compound formed from a number of connected pyrrole ring structures. For example, a tetrapyrrole is a compound with four pyrrole rings connected. Polypyrroles are conducting polymers of the rigid-rod polymer host family. Polypyrroles are also called **pyrrole blacks** or **polypyrrole blacks**.

Polythiophene

Polythiophenes (PTs) result from the polymerization of thiophenes, a sulfur heterocycle, that can become conducting when electrons are added or removed from the conjugated π-orbitals via doping.

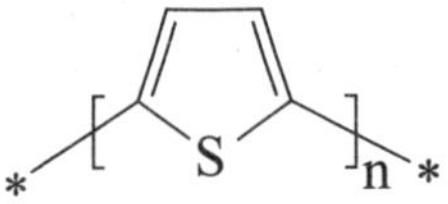

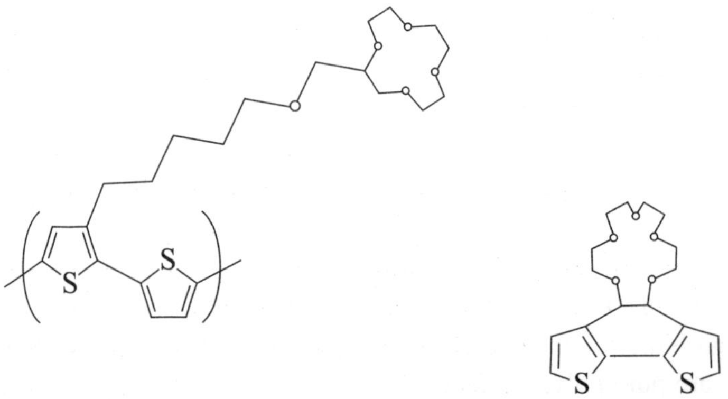

(a) (b)

Ionoselective PT with a) pendant crown ether functionalities
b) main-chain crown ether functionalities

Polythiophenes demonstrate interest-ing optical properties resulting from their conjugated backbone, as demonstrated by the fluorescence of a substituted polythiophene solution under UV irradiation. The optical properties of these materials respond to the environmental stimuli, with dramatic color shifts in response to changes in solvent, temperature, applied potential, and binding to other molecules.

The use of PTs as sensors responding to an analyte has also been the subject of intense research. In addition to biosensor applications, PTs can also be functionalized with synthetic receptors for detecting metal ions or chiral molecules as well. PTs with pendant and main-chain crown ether functionalities were reported in 1993 by the research groups of Bäuerle and Swager, respectively. Chiral PT was synthesized by Yashima and Goto.

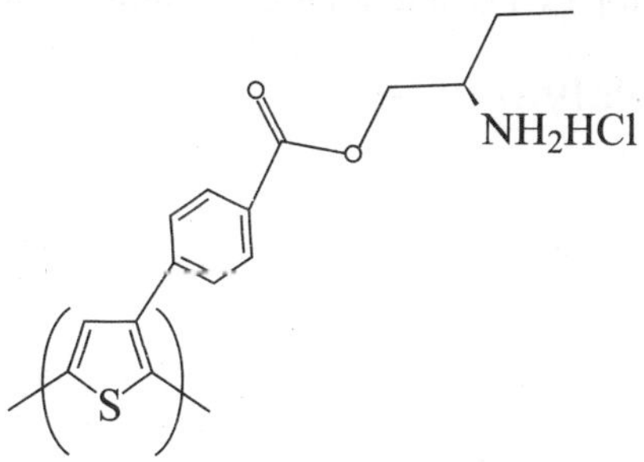

NH_2HCl

S

A number of applications have been proposed for conducting PTs, including field-effect transistors, electroluminescent devices, solar cells, photochemical resists, nonlinear optic devices, batteries, and diodes. In general, there are two categories of applications for conducting polymers. Static applications rely upon the intrinsic conductivity of the materials, combined with their ease of processing and material properties common to polymeric materials. Dynamic applications utilize changes in the conductive and optical properties, resulting either from application of electric potentials or from environmental stimuli.

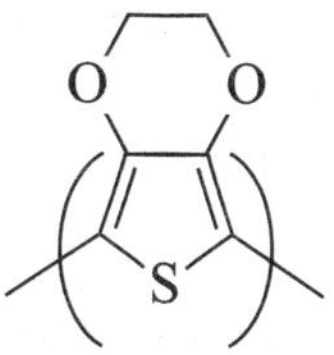

O O

S

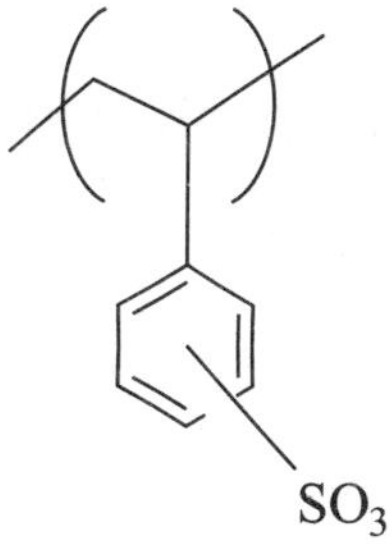

SO_3

Poly(3,4-ethylenedioxythiophene)-poly(styrene sulfonate) (PEDOT-PSS) has been extensively used as an antistatic coating (as packaging materials for electronic components, for example). PEDOT can also be used in dynamic applications where a potential is applied to a polymer film. The electrochromic properties of PEDOT are used to manufacture windows and mirrors which can become opaque or reflective upon the application of an electric potential. Widespread adoption of electrochromic windows could save billions of dollars per year in air conditioning costs.

Poly(3,4-ethylenedioxythiophene)

Poly(3,4-ethylenedioxythiophene) or PEDOT (or sometimes PEDT) is a conducting polymer based on 3,4-ethylenedioxylthiophene or EDOT monomer.

Advantages of this polymer are optical transparency in its conducting state, high stability and moderate band gap and low redox potential. A large disadvantage is poor solubility which is partly circumvented in the PEDOT:PSS composite. Pedot is a transparent conductor. These conductors are used for LCDs and solar cells, among others.

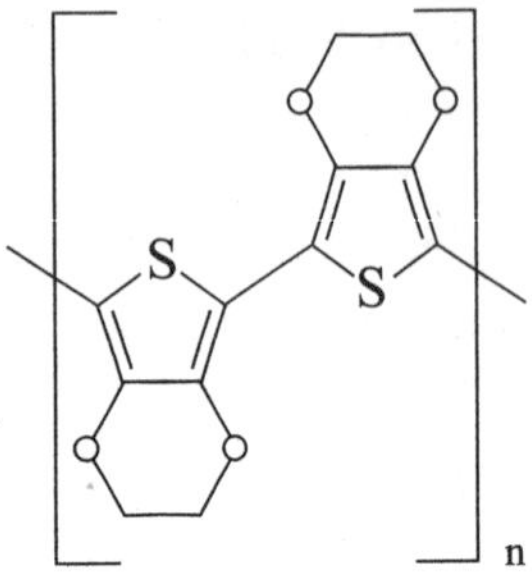

Polyaniline

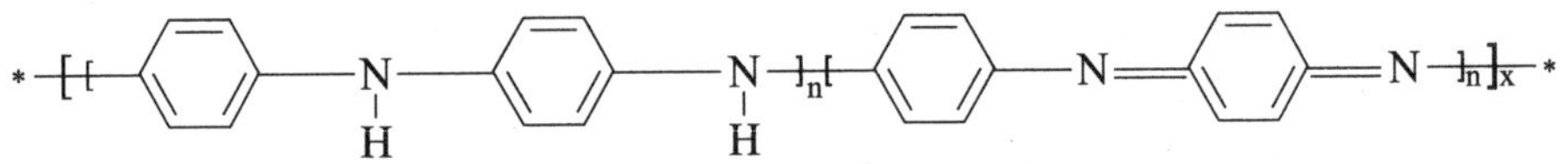

Polyaniline (PANI) is a conducting polymer of the semi-flexible rod polymer family. It was discovered in 1934 as aniline black. Polymerized from the aniline monomer, polyaniline can be found in one of the five distinct oxidation states- leucoemeraldine; protoemeralsine; emeraldine; nigraniline; pernigraniline. Emeraldine base is regarded as the most useful form of polyaniline due to its high stability at room temperature, compared to the easily oxidized leucoemeraldine and the easily degraded pernigraniline. In addition, the emeraldine base polyaniline can function as a semiconductor when doped by a protic acid.

Poly(p-phenylene oxide)

Poly(*p*-phenylene oxide) (PPO) can be used to prepare a conducting polymer of the semi-flexible rod polymer family. Oxidation or the use of dopants is used to convert the non-conductive PPO to a semiconductor.

Poly(p-phenylene sulfide)

Poly(*p*-phenylene sulfide) (PPS) is the precursor to a conducting polymer of the semi-flexible rod polymer family. One of the usual reactants in production of PPS is *p*-dichlorobenzene. The PPS may be converted to the semiconducting form by oxidation or use of dopants.

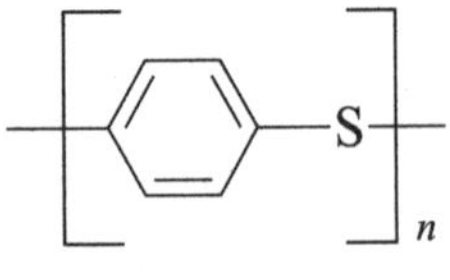

Poly(p-phenylene vinylene)

PPV, or Poly(*p*-phenylene vinylene) or Polyphenylene vinylene is a conducting polymer of the rigid-rod polymer host family. PPV is the only polymer of this type that has so far been successfully processed into a highly ordered crystalline thin film.

PPV is prepared by thermal processing of a precursor polymer. In presence of even a small amount of oxygen, singlet oxygen is formed during operation, by energy transfer from the excited polymer molecules to oxygen molecules.

These oxygen radicals then attack the structure of the polymer, leading to its degradation. Special precautions therefore have to be kept during manufacturing of PPV in order to prevent oxygen contamination. By careful control of the processing it is possible to influence the structure and morphology of the film. Polyphenylene vinylene is capable of electroluminescence, producing yellow-green light. It can be used for construction of organic light emitting diodes.

9.4 BIOPOLYMERS

9.4.1 Natural Biopolymers

Polymers which are found in nature or the synthetic polymers which are biodegradable or biocompatible are grouped in this class.

Natural macromolecules, e.g., proteins, cellulose and starch, are generally degraded in biological systems by hydrolysis followed by oxidation, and that is why many of the reported synthetic polymers also have hydrolysable linkages along the polymer chain, e.g., amine, enamine, ester, urea and urethane linkages are susceptible to biodegradation by microorganisms and hydrolytic enzymes.

Polymers with long repeating units are less likely to crystallize (means the structure of the polymer is irregular) and thus might be biodegradable, e.g., natural polymers contain long repeating units and irregular structure so it was thought that this property might be a factor in biodegradability.

Biopolymers can be used for a vast range of applications, including (but not limited to) adhesives, coatings, high performance fibres, plastics, cosmetics, pharmaceuticals, and medical and surgical devices. Natural biopolymers are carbohydrates, proteins, and nucleic Acids etc. Most common synthetic biopolymers include polylactic acid, polyacrylamide, polyethylene glycol, etc.

Polysaccharides

Polysaccharides are relatively complex carbohydrates. They are polymers made up of many monosaccharides joined together by glycosidic linkages. They are therefore very large, often branched, molecules. They tend to be amorphous, insoluble in water, and have no sweet taste. When all the constituent monosaccharides are of the same type they are termed homopolysaccharides; when more than one type of monosaccharide is present they are termed heteropolysaccharides. Polysaccharides have a general formula of $C_n(H_2O)_{n-1}$ where n is usually a large number between 200 and 2500.

Following is a list of some polysaccharides found in various natural resources:

Bacterial Polysaccharides

- Cellulose (bacterial)
- Xanthum
- Dextran
- Gellan
- Levan
- Curd lan
- Polygalactosamine

Fungal Polysaccharides

- Pullulan
- Elsinan
- Yeast glucans

Plant/algal Polysaccharides

- Starch
- Cellulose
- Agar
- Alginate
- Carrageenan
- Pectin
- Konjac
- Various gums (e.g., guar)

Animal Polysaccharides

- Chitan/chitosan
- Hyaluronic acid

Some commercially important polysaccharides are described here:

Starch

Biochemically, starch is a combination of two polymeric carbohydrates (polysaccharides) called amylose and amylopectin. Amylose is constituted by glucose monomer units joined to one another head-to-tail forming alpha-1,4 linkages. Amylopectin differs from amylose in the branching that occurs, with an alpha-1,6 linkage every 24-30 glucose monomer units. The overall structure of amylopectin is not that of a linear polysaccharide chain since two glucose units frequently form a branch point, so the result is the coiled molecule most suitable for storage in starch grains. Both amylopectin and amylose are polymers of glucose, and a typical starch polymer chain consists of around 2500 glucose molecules in their varied forms of polymerization. In general, starch has the formula $(C_6H_{10}O_5)_n$, where "n" denotes the total number of glucose monomer units.

Structurally, the starch forms clusters of linked linear polymers, where the alpha-1,4 linked chains form columns of glucose units which branch regularly at the alpha-1,6 links. The relative content of amylose and amylopectin varies between species, and between different cultivates of the same species. For example, high-amylose corn (maize) has starch consisting of about 85 per cent amylose, which is the linear constituent of starch, while waxy corn starch is more than 99 per cent amylopectin, or branched starch. The primary function of starch in plants, is to act as an energy storage molecule for the organism. In plants simple sugars are linked into starch molecules by specialized cellular organs called amyloplasts.

Starch is insoluble in water. It can be digested by hydrolysis, catalyzed by enzymes called amylases, which can break the glycosidic bonds between the 'alpha-glucose' components of the starch polysaccharide. Humans and other animals have amylases, so they can digest starch. Digestion of starch includes the cleavage of its molecules back into their constituent simple sugar units by the action of amylases. The resulting sugars are then processed further by enzymes (such as maltase) in the body, like other sugars in a diet.

Starch of all types are glucose polymers in which glucopyranose units are bonded by alpha-linkages. Amylose consists of a linear chain of several hundred glucose molecules. Amylopectin is a branched molecule made of several thousand glucose units.

Amylose is a linear polymer of glucose linked with mainly alpha-1,4 bonds. It can be made of several thousands of glucose units. It is one of the two components of starch, the other being amylopectin.

OH OH OH OH
O H O H O H O H
H H H H
O O O O
HO OH HO OH HO OH HO OH
n

Amylose

The alpha-1,4 bonds promote the formation of a helix structure. The number of repeated glucose subunits (*n*) can be many thousands. Amylose starch is less readily digested than amylopectin. However it takes up less space so it is preferred for storage in plants; that is how about 80 per cent of the starch in plants is stored. They can be digested by hydrolysis catalyzed by enzymes called amylases, which can break the alpha-linkages. Humans and other animals have amylases, so they can digest starch. Potato, rice, wheat, and maize are major sources of starch in a human diet.

Amylopectin is a highly branched polymer of glucose found in plants. It is one of the two components of starch, the other being amylose. It is insoluble in water. Glucose units are linked in a linear way with alpha-1,4 bonds. Branching takes place with alpha-1,6 bonds occurring every 24 to 30 glucose units. Starch is made of about 80 per cent amylopectin. Amylopectin is highly branched, being formed of 2,000 to 200,000 glucose units. Its counterpart in animals is glycogen which has the same composition and structure, but with more extensive branching that occurs every 8 to 12 glucose units.

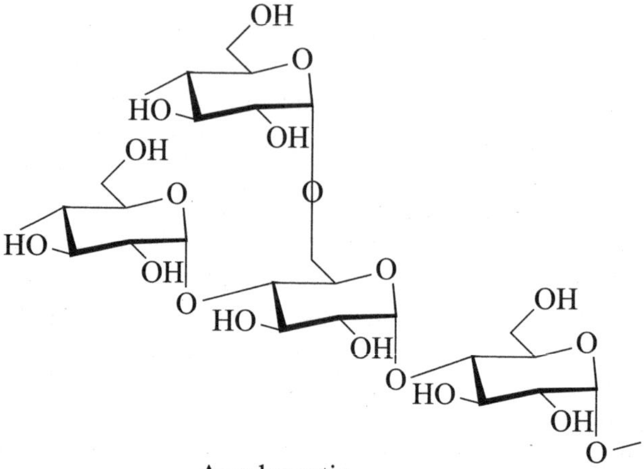

Amylopectin

As an additive for food processing, arrowroot, guar gum, locust bean, and tapioca are commonly used as well. Commonly used starch around the world are: arracacha, buckwheat, banana, barley, cassava, konjac, kudzu, oca, sago, sorghum, sweet potato, taro and yams. Edible beans, such as favas, lentils and peas, are also rich in starch. When a starch is pre-cooked it can then be used to thicken chilled foods. This is referred to on packaging as modified food starch. Agar, carrageenan, gelatins and pectins are used in very much the same way. Clothing starch or laundry starch is a liquid that is prepared by mixing a vegetable starch in water (earlier preparations also had to be boiled), and is used in the laundering of clothes.

Starch can be hydrolyzed into simpler carbohydrates (starch derivatives) by acids, various enzymes, or a combination of the two. The extent of conversion is typically quantified by dextrose equivalent (DE), which is roughly a fraction of the glycoside bonds in starch that have been broken. Food products made in this way include **maltodextrin**, a lightly hydrolyzed (DE 10–20) starch product used as a bland-tasting filler and thickener. Various **corn syrups** (DE 30–70), viscous solutions used as sweeteners and thickeners in many kinds of processed foods. Dextrose (DE 100), commercial glucose, are prepared by the complete hydrolysis of starch. High fructose syrup, made by treating dextrose solutions to the enzyme glucose isomerase, until a substantial fraction of the glucose has been converted to fructose. High fructose corn syrup is the principal sweetener used in sweetened beverages.

Glycogen

Glycogen is the storage form of glucose in animals. It is a branched polymer of glucose. Glycogen is found in the form of granules in the cytosol in many cell types. Hepatocytes (liver cells) have the highest concentration of it—up to 8 per cent of the fresh weight in well fed state, or 100–120 g in an adult—giving liver a distinctive, "starchy" taste. In the muscles, glycogen is found in a much lower concentration (1per cent of the muscle mass), but the total amount exceeds that in liver. Small amounts of glycogen are found in the kidneys, and even smaller amounts in certain glial cells in the brain and white blood cells.

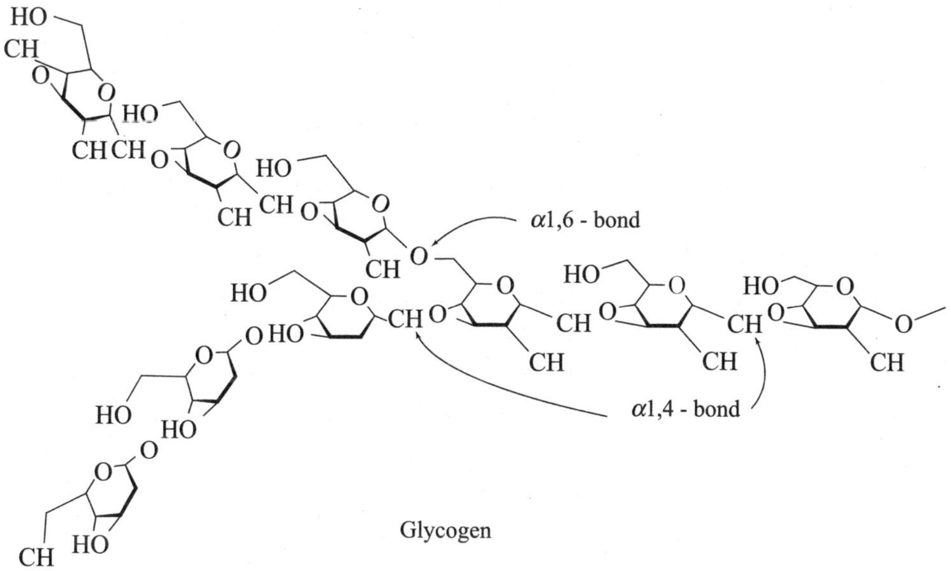

α1,6 - bond

α1,4 - bond

Glycogen

Glycogen can be broken down to form substrates for respiration, through the process of glycogenolysis. This involves the breaking of most of the C-O-C bonds between the glucose molecules by the addition of a phosphate, rather than water as in hydrolysis. This process yields phosphorylated glucose molecules, which can be metabolized with a saving of one ATP molecule.

Cellulose

Cellulose $(C_6H_{10}O_5)_n$ is a longchain polymeric polysaccharide carbohydrate. It forms the primary structural component of green plants. The primary cell wall of green plants is made primarily of cellulose. Cellulose is a polymer made with repeated glucose units bonded together by beta-linkages. The secondary cell wall contains cellulose with variable amounts of lignin. Lignin and cellulose, considered together, are termed lignocellulose, Wood, the most common biopolymer on earth, is largely cellulose and lignin, while paper and cotton are nearly pure cellulose.

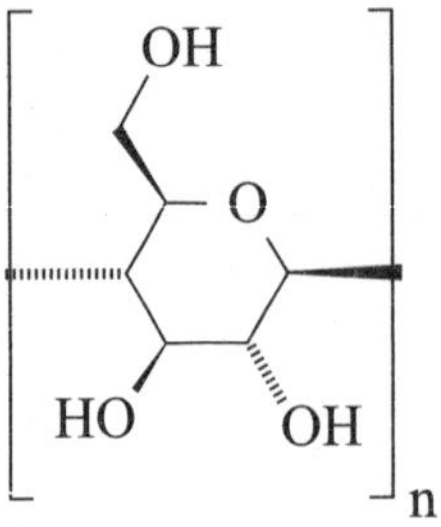

It is a versatile polymer and is found in nature in cotton seeds which contain 90–95 per cent cellulose. Wood contains 40–50 per cent, flax contains 80–85 per cent, hemp contains 70–80 per cent of cellulose.

Cellulose has a high melting point and it decomposes before it starts to melt. It is insoluble in most of the solvents. However, in some solvents, swelling may occur.

Cellulose

The hydroxyl groups of cellulose can be partially or fully reacted with various chemicals to provide derivates with useful properties. Cellulose esters and cellulose ethers are the most important commercial materials. In principle, though not always in current industrial practice, cellulosic polymers are renewable resources. Among the esters are cellulose acetate and triacetate, which are film- and fiber-forming materials that find a variety of uses. The inorganic ester nitrocellulose was initially used as an explosive and was an early film forming material. Ether derivatives include ethylcellulose, a water-insoluble commercial thermoplastic used in coatings, inks, binders, and controlled-release drug tablets; hydroxypropyl cellulose; carboxymethyl cellulose; hydroxypropyl methyl cellulose used as a viscosity modifier, geling agent, foaming agent and binding agent; hydroxyethyl methyl cellulose, used in production of cellulose films.

Cellulose is the most common natural polymer which is used as such in the form of cotton. The most common fibers made from cellulose are rayons.

Rayons

Rayon was the first fiber to be manufactured; it is produced from naturally occurring polymers and therefore it is not a synthetic fiber, but a manufactured regenerated cellulosic fiber. The fiber was sold as artificial silk until the name rayon was adopted in 1924. It is also known as viscose.

Rayon is a very versatile fiber and has the same comfort properties as other natural fibers and can imitate the feel and texture of silk, wool, cotton and linen. The fiber is easily dyed in a wide range of colors. Rayon fabrics are soft, smooth, cool, comfortable, and highly absorbent, but do not insulate body from heat making them ideal for use in hot and humid climates. The durability and appearance retention of regular rayon are low, especially when wet; also rayon has the lowest elastic recovery of any fiber. However, High Wet Modulus (HWM) rayon is much stronger and exhibits higher durability and appearance retention. Recommended care for regular rayon is dry-cleaning only, HWM rayon can also be machine washed.

Regular rayon (or viscose) is the most widely produced form of rayon. This method of rayon production has been utilized since the early 1900s and it has the ability to produce either filament or staple fibers. The process is as follows:

The cellulose is dissolved in caustic soda. This solution is then pressed between rollers to remove excess liquid. The pressed sheets are crumbled or shredded to produce what is known as white crumb. The white crumb is aged through exposure to oxygen. The aged white crumb is mixed with carbon disulfide in a process known as Xanthation. This process changes the chemical makeup of the cellulose mixture and the resulting product is now called yellow crumb. The yellow crumb is dissolved in a caustic solution to form viscose. The viscose is set to stand for a period of time, allowing it to ripen. After ripening, the viscose is filtered to remove any undissolved particles. Any bubbles of air are pressed from the viscose in a degassing process. The viscose solution is extruded through a spinneret, which resembles a shower head with many small holes. As the viscose exits the spinneret, it lands in a bath of sulfuric acid resulting in the formation of rayon filaments. The rayon filaments are stretched, known as drawing, to straighten out the fibers. The fibers are then washed to remove any residual chemicals. If filament fibers are desired the process ends here. The filaments are cut down when producing staple fibers.

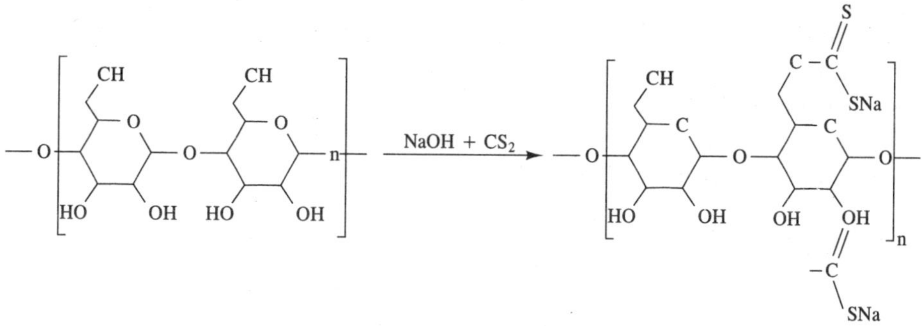

$NaOH + CS_2$

HWM is a modified version of viscose that has a greater strength when wet. It also has the ability to be mercerized like cotton. High Tenacity rayon is another modified version of viscose that has almost twice the strength of HWM. This type of rayon is typically used for industrial purposes such as tire cord. Cupramonium rayon has properties similar to viscose but during production, the cellulose is combined with copper and ammonia.

Some major rayon fiber uses include apparel (e.g., blouses, dresses, jackets, lingerie, linings, suits, ties), furnishings (e.g., bedspreads, blankets, window treatments, upholstery, slipcovers), industrial uses (e.g. medical surgery products, non-woven products, tire cord), and other uses (e.g. feminine hygiene products).

Carboxymethyl cellulose

Carboxymethyl cellulose, or CMC, is a cellulose derivative with carboxymethyl groups (—CH_2—COOH) bound to some of the hydroxyl groups of the glucopyranose monomers that make up the cellulose backbone.

It is synthesized by the alkali-catalyzed reaction of cellulose with chloroacetic acid. The polar (organic acid) carboxyl groups render the cellulose soluble and chemically reactive. The functional properties of CMC depend on the degree of substitution of the cellulose structure (i.e., how many of the hydroxyl groups have taken part in the substitution reaction), and also on the chain length of the cellulose backbone.

CMC is used in food science as a viscosity modifier or thickener, and to stabilize emulsions, for instance, in ice cream. It has E number E466. It is also a constituent of many non-food products, such as K-Y jelly, toothpaste, laxatives, diet pills, water-based paints, detergents, and various paper products. They have high viscosity, are not toxic, and are generally non-allergenic. CMC is used as a lubricant in non-volatile eye drops (artificial tears). Sometimes it is methylcellulose (MC) which is used, but its non-polar methyl groups ($-CH_3$) do not add any solubility or chemical reactivity to the base cellulose.

Celluloid

Celluloid is the name of a class of compounds created from nitrocellulose and camphor, plus dyes and other agents. Generally regarded to be the first thermoplastic, it was first created in 1856. Easily molded and shaped, there are suggestions that celluloid was first made as an ivory replacement. Celluloid is highly flammable and also easily decomposes, and is no longer widely used. Its most common use today is the table tennis ball.

Cellulose acetate

Cellulose acetate is the acetate ester of cellulose. Cellulose acetate is used as a film base in photography, and as a component in some adhesives; it is also used as a synthetic fiber.

Chitin

Chitin is one of the main components in the cell walls of fungi, the exoskeletons of insects and other arthropods, and in some other animals. It is a polysaccharide; it is constructed from units of acetylglucosamine (N-acetyl-D-glucose-2-amine). These are linked together in α-1,4 fashion (in a similar manner to the glucose units which form cellulose). In effect chitin may be described as cellulose with one hydroxyl group on each monomer replaced by an acetylamine group. This allows for increased hydrogen bonding between adjacent polymers, giving the polymer increased strength.

Chitin's properties as a tough and strong material make it favourable for use in surgical thread. Its biodegradability also means it wears away with time as the wound heals. Chitin also has some unusual properties that it accelerats healing of wounds in humans. It is closely related to chitosan (a more water-soluble derivative of chitin) and the immune system of plants and animals.

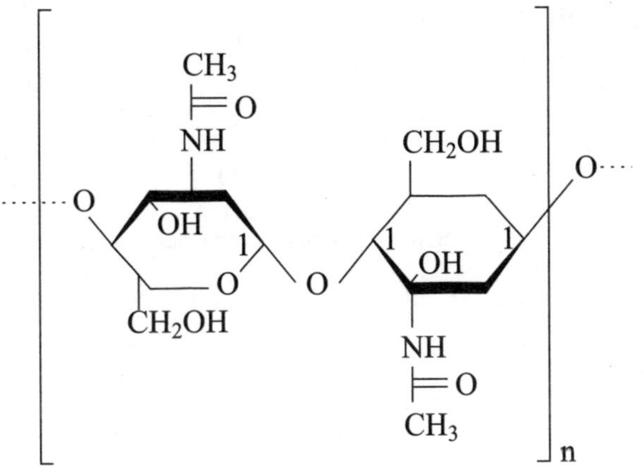

Chitin

Acidic polysaccharides

Acidic polysaccharides are those that contain carboxyl groups, phosphate groups and/or sulfuric ester groups. The most important acidic polysaccharide is alginic acid.

Alginic acid

Alginic acid (algine, alginate) is a viscous gum that is abundant in the cell walls of brown algae. Chemically, it is a linear copolymer with homopolymeric blocks of (1-4)-linked ß-D-mannuronate (M) and its C-5 epimer α-L-guluronate (G) residues, respectively, covalently linked together in different sequences or blocks.The monomers can appear in homopolymeric blocks of consecutive G-residues (G-blocks), consecutive M-residues (M-blocks), alternating M and G-residues (MG-blocks) or randomly organized blocks. The relative amount of each block type varies both with the origin of the alginate. Alternating blocks form the most flexible chains and are more soluble at lower pH than the other blocks. G-blocks form stiff chain elements, and two G-blocks of more than six residues each form stable cross-linked junctions with divalent cations (e.g. Ca^{2+}, Ba^{2+}, Sr^{2+} among others) leading to a three-dimensional gel network. At low pH, protonized alginates will form acidic gels. In these gels, it is mostly the homopolymeric blocks that form the junctions, where the stability of the gel is determined by the relative content of G-blocks.

Commercial varieties of alginate are extracted from seaweed, including the giant kelp *Macrocystis pyrifera*, *Ascophyllum nodosum* and various types of *Laminaria*. Alginate ranges from white to yellowish brown, and takes filamentous, grainy, granular, and powdered forms. It is insoluble in water and organic solvents, and dissolves slowly in basic solutions of sodium carbonate, sodium hydroxide and trisodium phosphate.

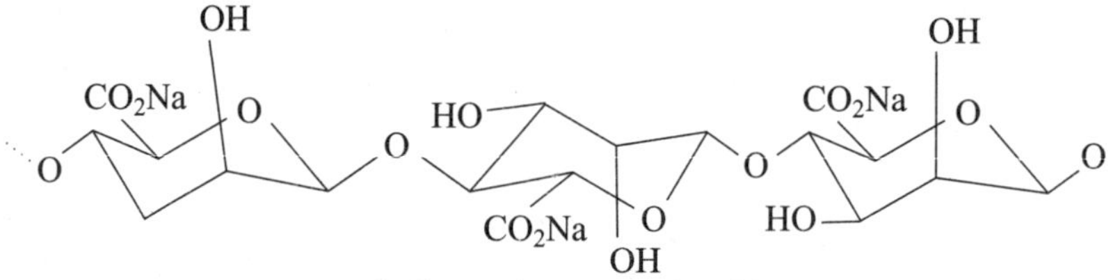

Sodium polymannuronic acid

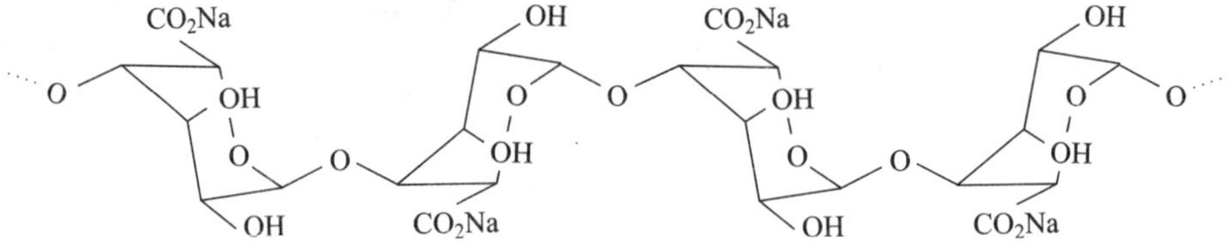

Sodium polygluronic acid

Alginate absorbs water quickly, which makes it useful as an additive in dehydrated products such as slimming aids, and in the manufacture of paper and textiles. It is also used for waterproofing and fireproofing fabrics, as a geling agent, for thickening drinks, ice-cream and cosmetics, and as a detoxifier that can absorb poisonous metals from the blood. Purified forms of alginate are used in antacid preparations. Alginate is used extensively as a mold-making material in dentistry and prosthetics, and in textiles. It is also used in the food industry, for thickening soups and jellies. Calcium alginate is used in burn dressings that promote healing and can be removed painlessly. Also, due to alginate's biocompatibility and simple gelation with divalent cations, it is widely used for cell immobilization and encapsulation.

Microbial Polysaccharides

Pathogenic bacteria commonly produce a thick, mucous-like, layer of polysaccharide. This capsule cloaks antigenic proteins on the bacterial surface that would otherwise provoke an immune response and thereby lead to the destruction of the bacteria. Capsular polysaccharides are water soluble, commonly acidic, and have molecular weights in the order of 100-1000 kDa. They are linear and consist of regularly repeating subunits of 1 ~ 6 monosaccharides. There is enormous structural diversity; nearly two hundred different polysaccharides are produced by *E. coli* alone. Mixtures of capsular polysaccharides, either conjugated or native are used as vaccines.

Bacteria and many other microbes, including fungi and algae, often secrete polysaccharides as an evolutionary adaptation to help them adhere to surfaces and to prevent them from drying out. Humans have developed some of these polysaccharides into useful products, including xanthan gum, dextran, gellan gum, and pullulan. Alginate is also produced by certain bacteria, notably *Azotobacter* species. Attempts to produce bacterial alginate have not yet been commercially successful.

Xanthan gum is a natural gum polysaccharide used as a food additive and rheology modifier. It is produced by a biotechnological process involving fermentation of glucose or sucrose by the *Xanthomonas campestris* bacterium.

Gellan gum is a polysaccharide produced by *Sphingomonas elodea*, a bacterium. Soluble in water, it is used primarily as an alternative to agar as a gelling agent in microbiological culture. Its largest advantages over agar are its almost perfect visual clarity and its strength; it is able to withstand 120 °C heat, making it especially useful in culturing thermophilic organisms.

Schizophyllan is a neutral extracellular polysaccharide produced by the fungus *Schizophyllan commune*, consisting of a 1,3-β-D-linked backbone of glucose residues with 1,6-β-D-glucosyl side groups. The non-degraded polysaccharide has a molecular weight of the order 10^6 Da. A chemically analogous polysaccharide, scleroglucan, is formed by the fungus *Sclerotium rolfsii*. Both polysaccharides share the chemical structure of the backbone with curdlan.

Kefiran is the clear or pale yellow slimy polysaccharide gel exuded by dairy kefir grains. Dairy kefir grains are symbiotic consortia of bacteria and yeasts which ferment mammal milk containing the complex sugar lactose. Kefiran has been shown to exhibit certain medicinal qualities, and it contributes to the organoleptic (mouthfeel) qualities of kefir.

Proteins

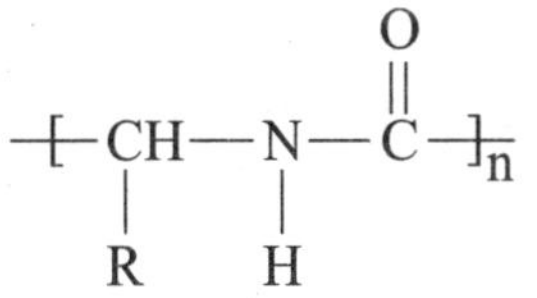

A protein is a naturally occurring polyamide. Proteins were the first examples of polyamides (a fancy word for nylon). Both share many common traits but they are very different in how they are made and in their physical properties. They are alike in that both contain amide linkages in the backbone. The differences between how nature does nylons and how man does it, is striking. Most of the nylons are made from molecules that have lots of CH_2 groups in them. In natural polyamides carbon is substituted with lots of different functional segments and groups.

$$\left[\underset{\displaystyle R}{\underset{|}{CH}} - \underset{\displaystyle H}{\underset{|}{N}} - \overset{\displaystyle O}{\overset{\|}{C}} \right]_n$$

Proteins are one of many types of natural polymers, and they are the most versatile so far. They can be catalysts called enzymes. One protein called hemoglobin, found in our blood, carries oxygen from lungs to cells. Another protein called collagen is a strong and tough material that makes up skin and fingernails, and even helps to hold your internal organs in place.

A protein is a complex, high-molecular-mass, organic compound that consists of amino acids joined by peptide bonds. Proteins are essential to the structure and function of all living cells and viruses. Different proteins perform a wide variety of biological functions. Some proteins are enzymes, which catalyze chemical reactions. Other proteins play structural or mechanical roles.

Proteins are a class of bio-macromolecules, alongside polysaccharides, lipids, and nucleic acids, that make up the primary constituents of biological organisms. Proteins are essentially polymers made up of a specific sequence of amino acids. The details of this sequence are stored in the code of a gene. Through the processes of transcription and translation, a cell reads the genetic information and uses it to construct the protein. In many cases, the resulting protein is then chemically altered (post-translational modification), before becoming functional. It is very common for proteins to work together to achieve a particular function, and often physically associate with one another to form a complex.

Some important proteins are listed below:

- Silks
- Collagen/gelatin
- Elastin (found in cows and pigs)
- Reslin adhesives
- Palamino acids
- Soy
- Zein from corn, wheat gluten, casein
- Serum albumin

Silks

Silk is a natural protein fiber that can be woven into textiles. It is obtained from the cocoon of silkworm larvae reared in captivity (sericulture). The shimmering appearance for which silk is prized comes from the fibers' triangular prism-like structure, which allows silk cloth to refract incoming light at different angles.

Spider silk is a fiber secreted by spiders. It is a remarkably strong material. Its tensile strength is comparable to that of high-grade steel. The proteins in the silk are complex molecules of amino acid. Silk is also used for items like parachutes, bicycle tires, comforter filling and artillery gunpowder bags. Silk undergoes a special manufacturing process to make it adequate for its use in surgery as non-absorbable sutures. Chinese doctors have also used it to make prosthetic arteries. Silk cloth is also used as a material to write on.

Zein

Zein is one of the most well understood plant proteins and has a variety of industrial and food uses. Historically it has been used in the manufacture of a wide variety of commercial products including coating paper cups, lining soda bottle caps, clothing fabric, buttons, adhesives, coatings and binders. Pure zein is clear, odorless, tasteless, hard, water-insoluble, and edible, making it invaluable in processed foods and pharmaceuticals, in competition with insect shellac. It is now used as a coating for candy, nuts, fruit, pills, and other encapsulated foods and drugs. In the United States it may be labeled as confectioner's glaze and used as a coating on bakery products or as vegetable protein. Zein can be further processed into resins and other bioplastic polymers which can be extruded or rolled into a variety of plastic products. With increasing environmental concerns about synthetic coatings (such as PFOA) and the currently higher prices of hydrocarbon based petrochemicals, there is increased focus on zein as a raw material for a variety of non-toxic and renewable polymer applications, particularly in paper industry applications. Other reasons for a renewed interest in zein include concern about the land-fill costs of plastics and consumer interest in the use of natural substances. There are also a number of new food industry potential applications.

Enzymes

Enzymes are one of the key types of polypeptides and are crucial to life on the earth. Enzymes are the catalysts that do specific jobs. In fact, most of the times each enzyme does only one type of job or makes only one kind of molecule. This means that there have to be lots of different enzymes, all made of different combinations of amino acids joined in unique ways in polypeptides, to do all the jobs that any living organism needs to be done.

DNA and RNA

RNA and DNA contain polymer backbones which are based on sugar units. This makes them polysaccharides, although in the case of RNA and DNA, there are well ordered groups attached to the sugar units that give these polymers their unique capabilities.

Polyisoprene is also a natural polymer but it is discussed under subsection 9.2. Similarly modified cellulose, although not purely natural, are discussed in the same section only.

9.4.2. Synthetic Biopolymers

Biodegradable polymers are used in medical field (sutures, surgical implant, and controlled release formulations of drug) and agricultural chemicals. If somchow we make only such polymers which are biodegradable then the disposal of polymeric materials will not be a problem. Examples of biodegradable synthetic polymers are

(a) polymer with hydrolyzable backbone:

1. PGA
2. Poly (α-hydroxy acid)
3. Poly (glycolic acid)
4. Poly (L-lactic acid) PLLA
5. Copolymer of PET and PLLA
6. Polyamide urethane
7. Polyamides: (i) hydrozylated and methylated nylons, (ii) benzylated nylons

(b) polymers with carbon backbones:

1. Poly(vinyl alcohol)
2. Poly(vinyl acetate)
3. Poly(alkyl acrylates)

Incorporation of groups which are susceptible to oxidation might improve their biodegradability.

Polylactic acid

Polylactic acid or Polylactide (PLA) is a biodegradable, thermoplastic, aliphatic polyester derived from lactic acid. It can be easily produced in a high molecular weight form through ring-opening polymerization using a stannous octoate catalyst, but for laboratory demonstrations tin(II) chloride is often employed.

$$\text{Lactide} \xrightarrow[\text{Heat}]{\text{Catalyst +}} \left[-O-\underset{CH_3}{\underset{|}{CH}}-\overset{O}{\overset{\|}{C}}- \right]_n -O-\underset{CH_3}{\underset{|}{CH}}-\overset{O}{\overset{\|}{C}}-$$

Lactide Polylactide

Due to the chiral nature of lactic acid several distinct forms of polylactide exist: poly-L-lactide (PLLA) is the product resulting from polymerization of lactic acid in the L form. PLLA has a crystallinity around 37 per cent, a glass transition temperature between 50–80 °C and a melting temperature between 173–178 °C. The polymerization of a mixture of both L and D forms of lactic acid leads to the synthesis of poly-DL-lactide (PDLLA) which is not crystalline but amorphous. Polylactic acid can be processed like most thermoplastics into fiber (for example, using conventional melt-spinning processes) and film. PLA is currently used in a number of biomedical applications, such as sutures, dialysis media and drug delivery devices, but it is also evaluated as a material for tissue engineering. Being biodegradable it can also be employed in the preparation of bioplastic, useful for producing loose-fill packaging, compost bags, food packaging and disposable tableware. In form of fibers and non-woven textiles PLA also has many potential uses, for example, as upholstery, disposable garments, feminine hygiene products and nappies.

PLA is particularly attractive as a sustainable alternative to petrochemical derived products, since the lactate from which it is ultimately produced can be derived from the fermentation of agricultural by-products such as corn starch or other starch-rich substances like maize, sugar or wheat. PLA is more expensive than many petroleum derived commodity plastics, but its price has been falling as more production comes online. The degree to which the price will fall, and the degree to which PLA will be able to compete with non-sustainable petroleum-derived polymers, is uncertain.

Polycaprolactone

Polycaprolactone (PCL) is a biodegradable polyester with a low melting point of around 60°C and a glass transition temperature of about –60°C. PCL can be prepared by ring opening polymerization of ε-caprolactone using a catalyst such as stannous octanoate.

$$\text{ε-Caprolactone} \xrightarrow[\text{+ Heat}]{\text{Catalyst}} \left[-O-(CH_2)_5-\overset{O}{\overset{\|}{C}}- \right]_n$$

ε-Caprolactone PolyCaprolactone

This polymer is often used as an additive for resins to improve their processing characteristics and their end use properties (e.g., impact resistance). Being compatible with a range of other materials, PCL can be mixed with starch to lower its cost and increase biodegradability or it can be added as a polymeric plasticizer to PVC. Polycaprolactone is also used for splinting, modelling, and as a feedstock for prototype. Fused Deposition Modeling systems such as RepRap.

Biomedical Applications: PCL is degraded by hydrolysis of its ester linkages in physiological conditions (such as in the human body) and has therefore received a great deal of attention for use as an implantable biomaterial. In particular it is especially interesting for the preparation of long term implantable devices, owing to its degradation which is even slower than that of polylactide. PCL is a Food and Drug Administration (FDA) approved material that is used in the human body as (for example) a drug delivery device, suture, adhesion barrier and is being investigated as a scaffold for tissue repair via tissue engineering. A variety of drugs have been encapsulated within PCL beads for controlled release and targeted drug delivery.

Polyglycolide

Polyglycolide or polyglycolic acid (PGA) is a biodegradable, thermoplastic polymer and the simplest linear, aliphatic polyester. It can be prepared starting from glycolic acid by means of polycondensation or ring-opening polymerization. PGA has been known since 1954 as a tough fiber forming polymer, however, owing to its hydrolytic instability its use has initially been limited. Currently polyglycolide and its copolymers poly(lactic-co-glycolic acid) with lactic acid, poly(glycolide-co-caprolactone) with ε-caprolactone and poly(glycolide-co-trimethylene carbonate) with trimethylene carbonate are widely used as a material for the synthesis of absorbable sutures and are being evaluated in the biomedical field.

Polyglycolide has a glass transition temperature between 35–40 °C and its melting point is reported to be in the range of 225–230 °C. PGA also exhibits an elevated degree of crystallinity, around 45–55per cent thus resulting insoluble in water. The solubility of this polyester is somewhat unique, in that its high molecular weight form is insoluble in almost all common organic solvents (acetone, dichloromethane, chloroform, ethyl acetate, tetrahydrofuran), while low molecular weight oligomers sufficiently differ in their physical properties to be more soluble. However, polyglycolide is soluble in highly fluorinated solvents like hexafluoroisopropanol (HFIP) and hexafluoroacetone sesquihydrate, that can be used to prepare solutions of the high MW polymer for melt spinning and film preparation. Fibers of PGA exhibit high strength and modulus (7 GPa) and are particularly stiff.

$$\text{Glycolide} \xrightarrow{\text{Catalyst + Heat}} \left[-O-CH_2-\overset{O}{\overset{\|}{C}}- \right]_n O-CH_2-\overset{O}{\overset{\|}{C}}-$$

Glycolide Polyglycolide

Polyglycolide can be obtained through several different processes starting with different materials—polycondensation of glycolic acid; ring-opening polymerization of glycolide; solid-state polycondensation of halogenoacetates; acid catalyzed reaction of carbon monoxide and formaldehyde.

Because polyglycolide gives strongs fibers and degrades into water soluble monomers, sutures made with this polymer have found use in certain surgical applications, since there is no need for further medical attention to remove them. Implantable medical devices have been produced with PGA as well, including anastomosis rings, pins, rods, plates and screws. The traditional role of PGA as a biodegradable suture material has led to its evaluation in other biomedical fields, such as tissue engineering or controlled drug delivery. Tissue engineering scaffolds made with polyglycolide have been produced following different approaches, but generally most of these are obtained through textile technologies in the form of non-woven meshes.

Cyclodextrin

Cyclodextrins (sometimes called cycloamyloses) make up a family of cyclic oligosaccharides, composed of 5 or more α-D-glucopyranoside units linked 1, 4 as in amylose (a fragment of starch). The 5-membered macrocycle is not natural. Recently, the largest well-characterized cyclodextrin contains 32 1,4-anhydroglucopyranoside units, while as a poorly characterized mixture, even at least 150-membered cyclic oligosaccharides are also known.

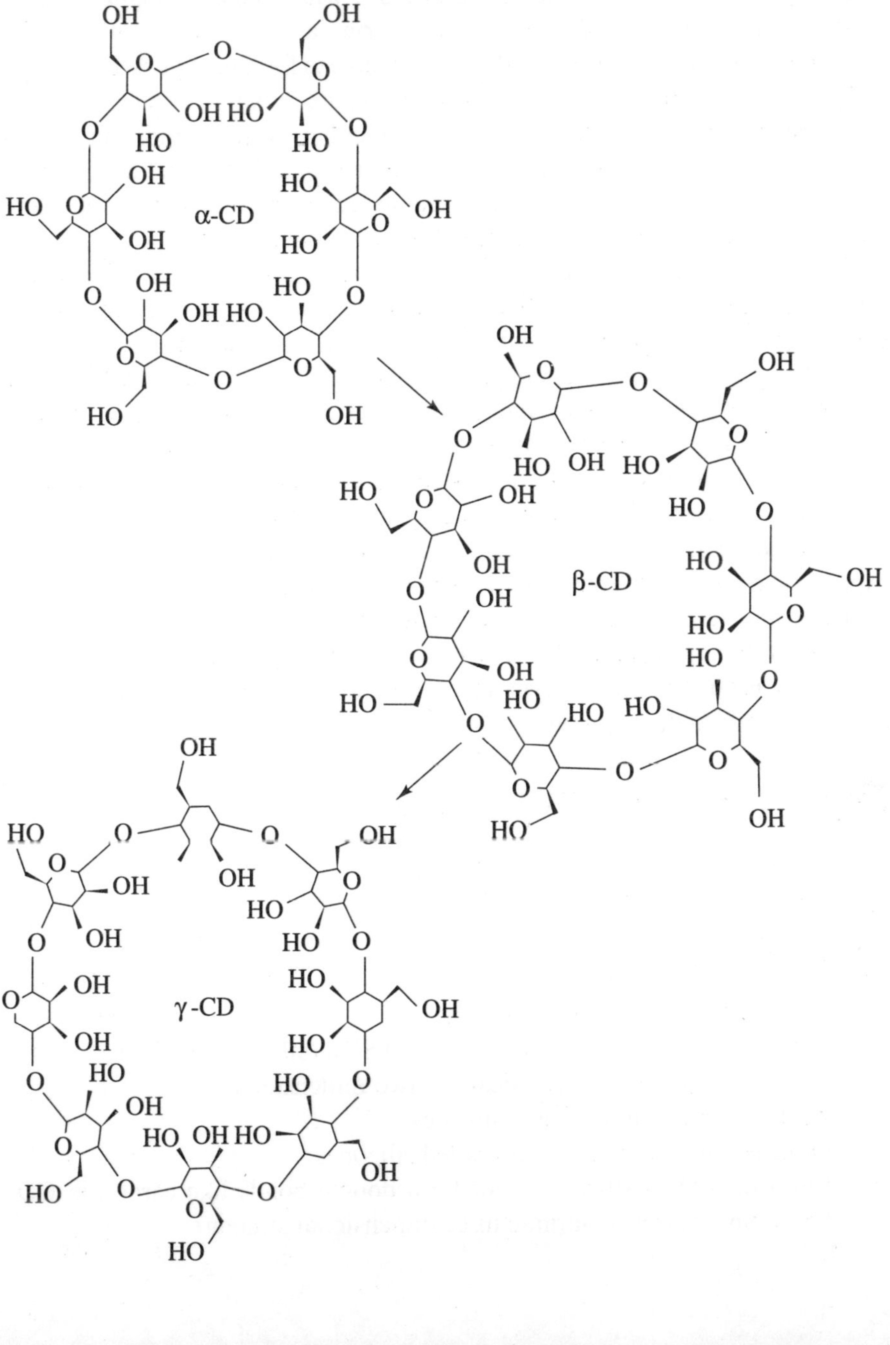

Typical cyclodextrins contain a number of glucose monomers ranging from six to eight units in a ring, thus denoting:

α-cyclodextrin: six membered ring molecule
β-cyclodextrin: seven membered ring molecule
γ-cyclodextrin: eight membered ring molecule

Cyclodextrins are produced from starch by means of enzymatic conversion. The production of cyclodextrins is relatively simple and involves treatment of ordinary starch with a set of easily available enzymes.

Over the last few years they have found a wide range of applications in food, pharmaceutical and chemical industries as well as agriculture and environmental engineering. Cyclodextrins are able to form host-guest complexes with hydrophobic molecules given the unique nature imparted by their structure. As a result, these molecules have found a number of applications in a wide range of fields. Other than the above mentioned pharmaceutical applications for drug release, cyclodextrins can be employed in environmental protection: these molecules can effectively immobilise inside their rings, toxic compounds like trichloroethane or heavy metals, or can form complexes with stable substances like trichlorfon (an organophosphorus insecticide) or sewage sludge, enhacing their decomposition. In the food industry, cyclodextrins are employed for the preparation of cholesterol free products: the bulky and hydrophobic cholesterol molecule is easily lodged inside cyclodextrin rings that are then removed, leaving behind a low fat food. Other food applications further include the ability to stabilize volatile or unstable compounds and the reduction of unwanted tastes and odour. The strong ability of complexing fragrances can also be used for another purpose: first dry, solid cyclodextrin microparticles are exposed to a controlled contact with fumes of active compounds, then they are added to fabric or paper products. Such devices are capable of releasing fragrances during ironing or when heated by human body. Such a device commonly used is a typical dryer sheet. The heat from clothes dryer releases the fragrance into the clothing.

9.5 INORGANIC POLYMERS

Polymers whose backbone chains are made of atoms other than carbon atoms are termed as inorganic polymers.

Silica Polymers

Following are the types of silica polymers:

(1) Asbestos : (a) infinite helical chains of SiO_3 (silicone trioxide)
(b) one sheet silicates – two dimensional
(2) Zeolites – three dimensional silicates
(3) Organosilicon polymers – are called silicones
(4) Silica or SiO_2 – silicon cannot form double bonds like carbon in CO_2 hence SiO_2 forms an infinite three dimensional structure.

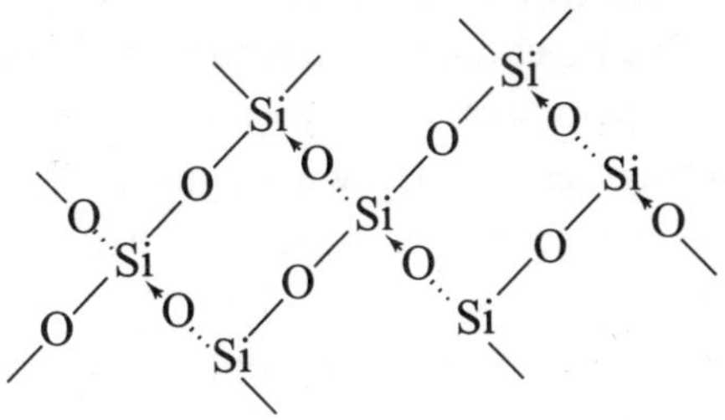

SiO_2 in its crystalline form–quartz.

Thus SiO_2 is a high temperature melting solid which exist in three forms:

At low temperature	α-quartz	α-tridymite	α-cristosalite
At high temperature	β-quartz	β-tridymite	β-cristosalite

These six forms are inter-convertible.

Polysiloxanes

Silicones

Silicones, or polysiloxanes, are inorganic-organic polymers with the chemical formula $[R_2SiO]_n$, where R = organic groups such as methyl, ethyl, and phenyl. These materials consist of an inorganic silicon-oxygen backbone (...-Si-O-Si-O-Si-O-...) with organic side groups attached to the silicon atoms, which are four-coordinate. In some cases organic side groups can be used to link two or more of these -Si-O- backbones together. By varying the -Si-O- chain lengths, side groups, and crosslinking, silicones can be synthesized with a wide variety of properties and compositions. They can vary in consistency from liquid to gel to rubber to hard plastic. The most common type is linear polydimethylsiloxane or PDMS. The second largest group of silicone materials is based on silicone resins, which are formed by branched and cage-like oligosiloxanes.

Silicones are odorless, colorless, water resistant, chemical resistant, oxidation resistant, stable at high temperature, and have weak forces of attraction, low surface tension, low freezing points and do not conduct electricity. They have many uses, such as lubricants, adhesives, sealants, gaskets, breast implants, pressure compensating diaphragms for drip irrigation emitters, dishware, Silly Putty, and many other products. Due to their thermal stability and relatively high melting and boiling points, silicones are often used where organic polymers are not applicable. Their unreactivity generally makes them non-toxic. Silicones are the most common polysiloxanes.

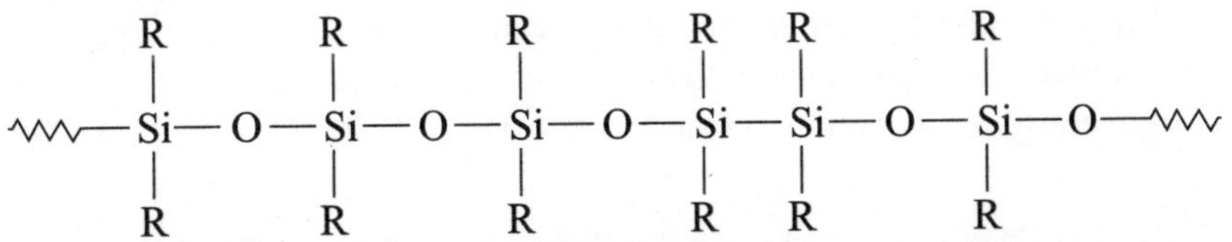

Silicone

The bond between silicon and oxygen is very strong, but very flexible. So silicones can stand high temperatures without decomposing, but they have very low glass transition temperatures. Silicone polymers are thermally stable and used in surface coatings especially where water repelling and chemical resistance are important considerations. Laminates made out of silicone polymers can withstand high temperatures.

Using foaming agents silicone foams are obtained. These are light weight so used in aeroplanes and missiles. Like rubber, the silicons rubbers can also be vulcanized with the help of peroxide initiators.

Glass

Glass could be considered as a highly crosslinked polysiloxane.

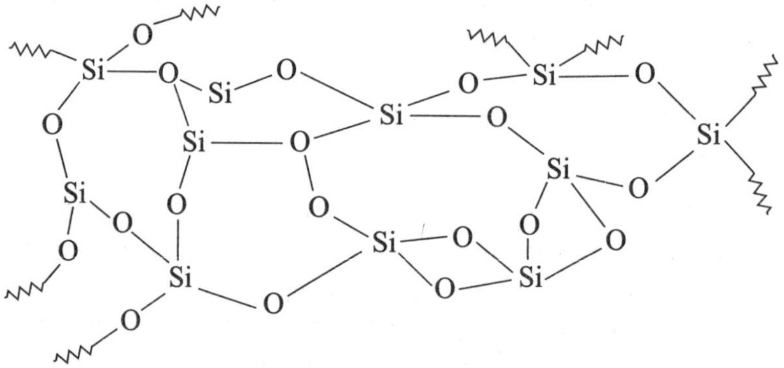

Glass

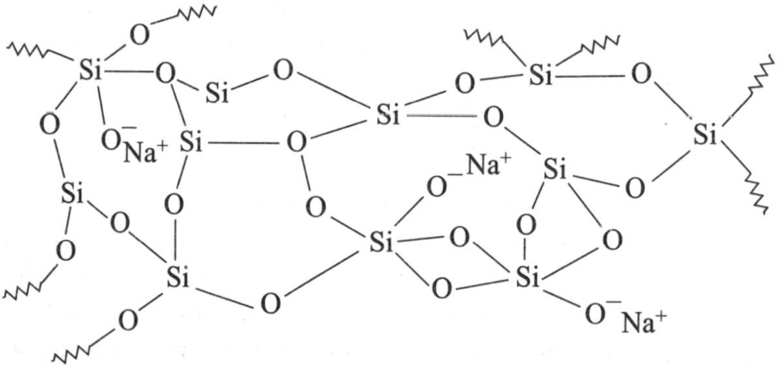

Tougher Glass

Polydimethylsiloxane

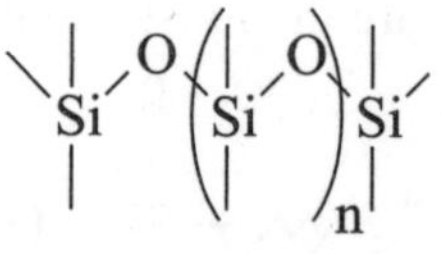

Polydimethylsiloxane (PDMS) is the most widely used silicon-based organic polymer, and is particularly known for its unusual rheological properties. It is optically clear, and is generally considered to be inert, non-toxic and nonflammable. PDMS is occasionally called dimethicone. It is one of several types of silicone oil. It is the primary component in Silly Putty, to which PDMS imparts its characteristic viscoelastic properties. The rubbery, vinegary-smelling silicone caulks, adhesives, and aquarium sealants are also well-known. PDMS is also used as a component in silicone grease and other silicone based lubricants, as well as in defoaming agents, damping fluids, heat transfer fluids, cosmetics, hair conditioner and other applications.Many people are indirectly familiar with PDMS. PDMS has also been used as a filler fluid in breast implants, although this practice has decreased somewhat, due to safety concerns. It continues to be used in knuckle replacement implants, with good results.

Silicone rubber

These polymers have a backbone of silicon oxygen linkages, the same bond that is found in quartz, glass and sand. This is one of the main reasons that silicone rubber has such excellent high temperature properties. Organic rubber has a carbon to carbon backbone. This can leave them susceptible to ozone, UV, heat and other aging factors that silicone rubber can withstand well. This is why it is the material of choice in many extreme environments.

Silicone rubber offers incredible resistance to extreme temperatures, being able to operate normally from –100 °C to +300 °C. In such conditions the tensile strength, elongation, tear strength and compression set can be far superior to conventional rubbers. There are also many specialist grades of silicone rubber that offer the following qualities: steam resistance, metal detectable, glow in the dark, electrically conductive, chemical/oil/acid/gas resistance, low smoke emission and flame retardent.

Once milled and colored the silicone rubber can be extruded into tubes, strips, solid cord or custom profiles within the size restrictions of the manufacturer. Cord can be joined to make "O" rings and extruded profiles can also be joined to make up seals.

Silicone rubber can also be molded into custom shapes and designs. Heat is required to vulcanize (set) the silicone rubber, this is normally carried out in a two stage process at the point of manufacture into the desired shape, and then in a prolonged post-cure process.

Silicone resin

Silicone resins are a type of silicone material which is formed by branched, cage-like oligosiloxanes with the general formula of $R_n\mathrm{SiX}_m\mathrm{O}_y$, where R is a nonreactive substituent, usually Me or Ph, and X is a functional group H, OH, Cl or OR. These groups are further condensed in many applications, to give highly crosslinked, insoluble polysiloxane networks.

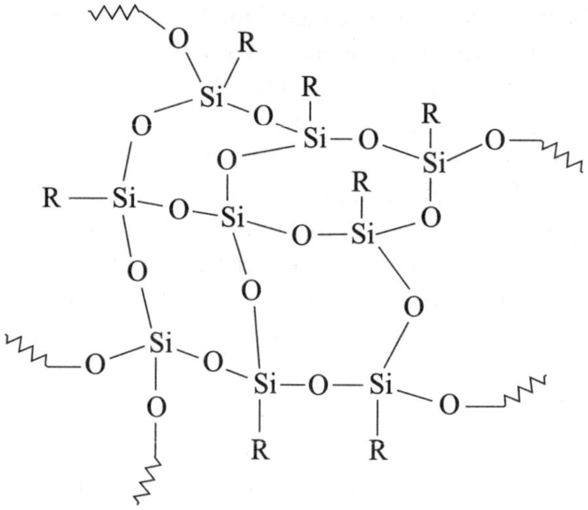

R = Me, OH, H

When R is methyl, the four possible functional siloxane monomeric units are described as follows:

M stands for Me_3SiO; D for Me_2SiO_2; T for $MeSiO_3$ and Q for SiO_4.

Note that a network of only Q groups becomes fused quartz.

The most abundant silicone resins are built of D and T units (DT resins) or from M and Q units (MQ resins), however, many other combinations (MDT, MTQ, QDT) are also used in industry.

Silicone resins represent a broad range of products. Materials of molecular weight in the range of 1000–10 000 are very useful in pressure sensitive adhesives, silicone rubbers, coatings and additives.

Silicone resins are prepared by hydrolytic condensation of various silicone precursors. In early processes of preparation of silicone resins sodium silicate and various chlorosilanes were used as starting materials. Although the starting materials were the least expensive ones (something typical for industry), structural control of the product was very difficult. More recently, a less reactive tetraethoxysilane (TEOS) or ethyl polysilicate and various disiloxanes are used as the starting materials. The silicon resins are obtained as follows:

(i) Dimethyl-silicon dichloride on hydrolysis (easy hydrolysis) with water gives the corresponding hydroxy compound (called silenols), which on self-condensation give the silicon resin.

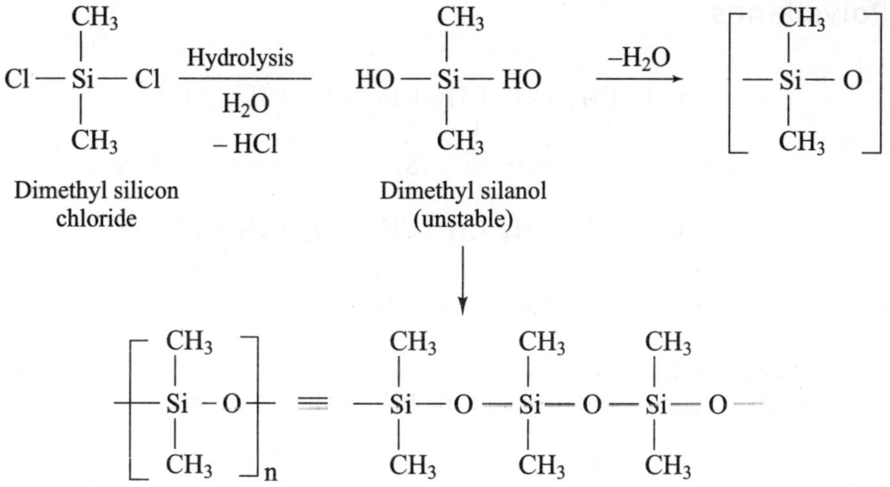

$$\underset{\text{Dimethyl silicon chloride}}{Cl-\overset{\overset{CH_3}{|}}{\underset{\underset{CH_3}{|}}{Si}}-Cl} \xrightarrow[\substack{H_2O \\ -HCl}]{\text{Hydrolysis}} \underset{\text{Dimethyl silanol (unstable)}}{HO-\overset{\overset{CH_3}{|}}{\underset{\underset{CH_3}{|}}{Si}}-HO} \xrightarrow{-H_2O} \left[-\overset{\overset{CH_3}{|}}{\underset{\underset{CH_3}{|}}{Si}}-O\right]$$

$$\downarrow$$

$$\left[-\overset{\overset{CH_3}{|}}{\underset{\underset{CH_3}{|}}{Si}}-O-\right]_n \equiv -\overset{\overset{CH_3}{|}}{\underset{\underset{CH_3}{|}}{Si}}-O-\overset{\overset{CH_3}{|}}{\underset{\underset{CH_3}{|}}{Si}}-O-\overset{\overset{CH_3}{|}}{\underset{\underset{CH_3}{|}}{Si}}-O-$$

Dimethyl silicon polymer

By this process very long linear polymers are obtained. (ii) Monomethyl silicon chloride on reaction with water gives the corresponding hydroxy compounds (silanol) which are unstable and polymerise to give cross linked polymers.

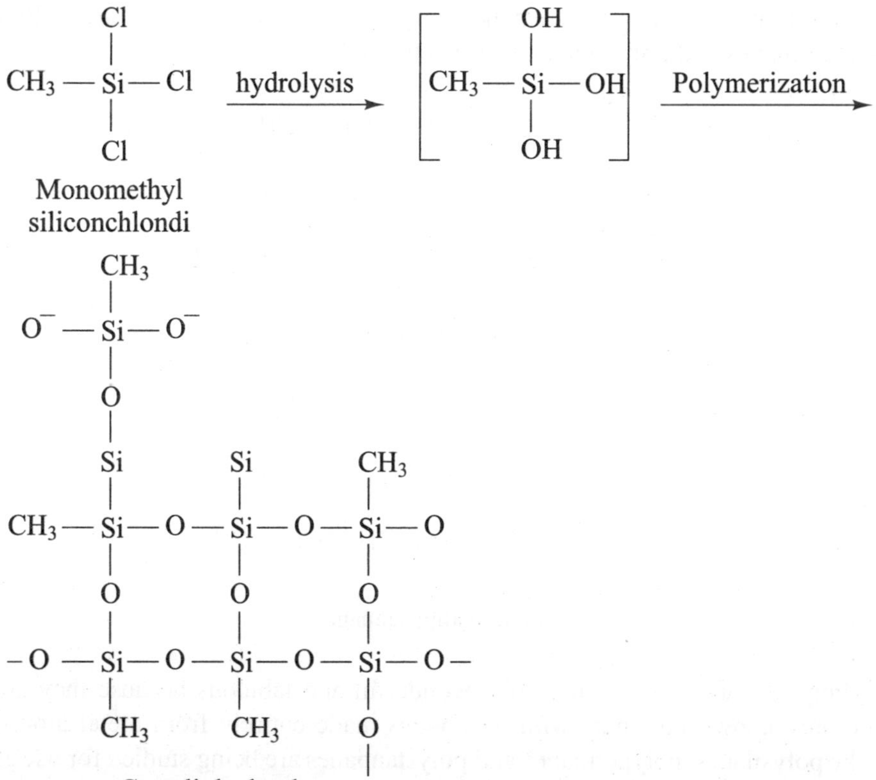

$$\underset{\text{Monomethyl siliconchlondi}}{CH_3-\overset{\overset{Cl}{|}}{\underset{\underset{Cl}{|}}{Si}}-Cl} \xrightarrow{\text{hydrolysis}} \left[CH_3-\overset{\overset{OH}{|}}{\underset{\underset{OH}{|}}{Si}}-OH\right] \xrightarrow{\text{Polymerization}}$$

$$\begin{array}{ccccc}
 & & & & \\
O^- - & \overset{\overset{CH_3}{|}}{Si} & -O^- & & \\
 & | & & & \\
 & O & & & \\
 & | & & & \\
 & Si & & Si & & CH_3 \\
 & | & & | & & | \\
CH_3- & Si & -O- & Si & -O- & Si-O \\
 & | & & | & & | \\
 & O & & O & & O \\
 & | & & | & & | \\
-O- & Si & -O- & Si & -O- & Si-O- \\
 & | & & | & & | \\
 & CH_3 & & CH_3 & & O \\
 & & & & & |
\end{array}$$

Crosslinked polymer

Polysilanes

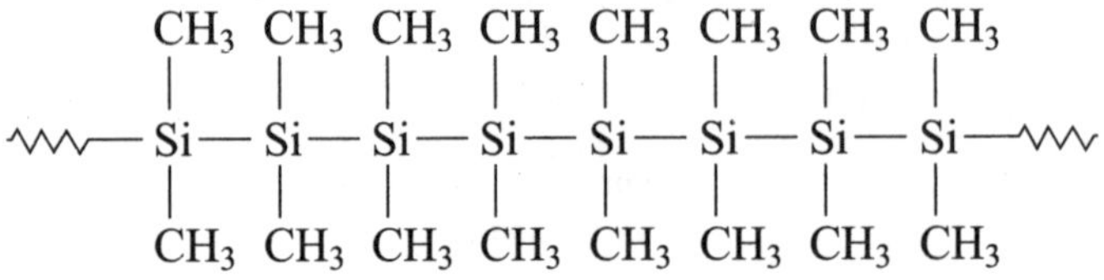

$$\begin{array}{ccccccccccccccccc} & CH_3 & & CH_3 & & CH_3 & & CH_3 & & CH_3 & & CH_3 & & CH_3 & & CH_3 & \\ & | & & | & & | & & | & & | & & | & & | & & | & \\ \sim\!\sim - & Si & - & Si & - & Si & - & Si & - & Si & - & Si & - & Si & - & Si & - \sim\!\sim \\ & | & & | & & | & & | & & | & & | & & | & & | & \\ & CH_3 & & CH_3 & & CH_3 & & CH_3 & & CH_3 & & CH_3 & & CH_3 & & CH_3 & \end{array}$$

Polydimethylsilane

Polysilanes form crystals that are so strong that nothing can dissolve them. It is made by reacting sodium metal with dichlorodimethylsilane:

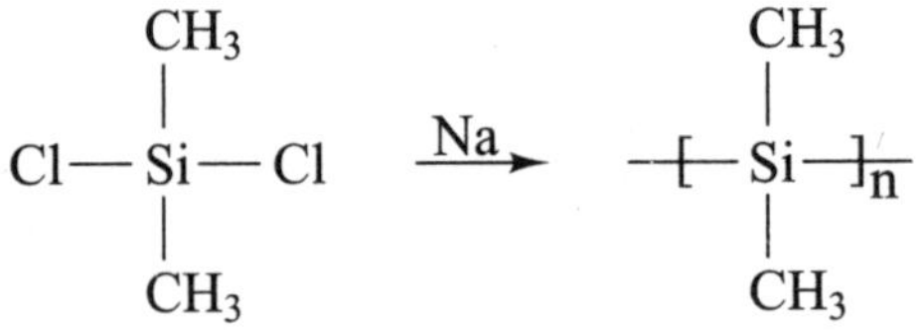

$$\begin{array}{ccc} & CH_3 & \\ & | & \\ Cl- & Si & -Cl \\ & | & \\ & CH_3 & \end{array} \xrightarrow{Na} \left[\begin{array}{c} CH_3 \\ | \\ -Si- \\ | \\ CH_3 \end{array} \right]_n$$

Polygermanes and Polystannanes

If silicon can make long polymer chains, the other elements in Group IV like germanium and tin can also make polymers. These polymers are called polygermanes and polystannanes, respectively.

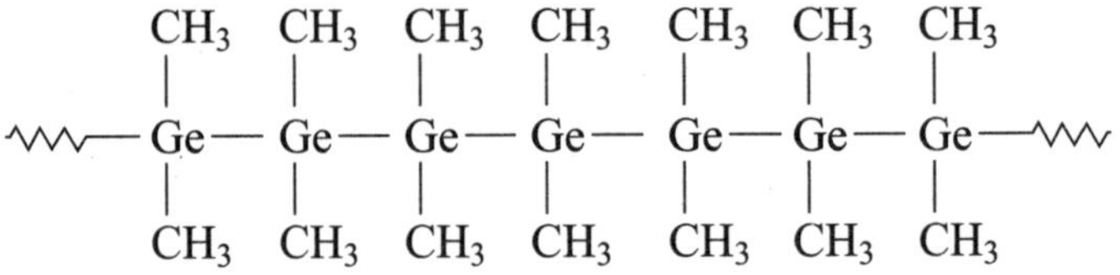

$$\begin{array}{ccccccccccccccc} & CH_3 & & CH_3 & & CH_3 & & CH_3 & & CH_3 & & CH_3 & & CH_3 & \\ & | & & | & & | & & | & & | & & | & & | & \\ \sim\!\sim - & Ge & - & Ge & - & Ge & - & Ge & - & Ge & - & Ge & - & Ge & - \sim\!\sim \\ & | & & | & & | & & | & & | & & | & & | & \\ & CH_3 & & CH_3 & & CH_3 & & CH_3 & & CH_3 & & CH_3 & & CH_3 & \end{array}$$

polydimethylgermane

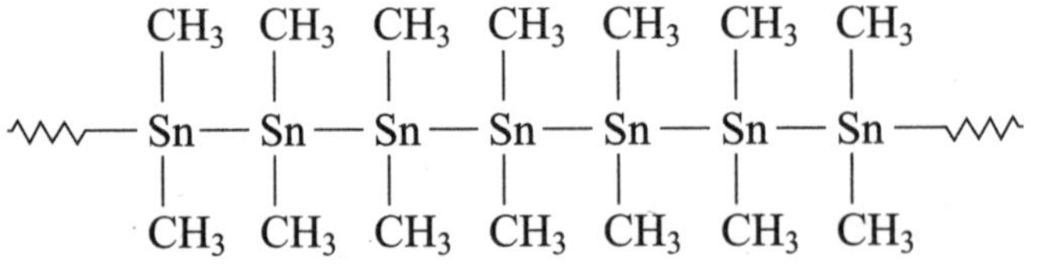

$$\begin{array}{ccccccccccccccc} & CH_3 & & CH_3 & & CH_3 & & CH_3 & & CH_3 & & CH_3 & & CH_3 & \\ & | & & | & & | & & | & & | & & | & & | & \\ \sim\!\sim - & Sn & - & Sn & - & Sn & - & Sn & - & Sn & - & Sn & - & Sn & - \sim\!\sim \\ & | & & | & & | & & | & & | & & | & & | & \\ & CH_3 & & CH_3 & & CH_3 & & CH_3 & & CH_3 & & CH_3 & & CH_3 & \end{array}$$

polydimethylstannane

Polystannanes are unique, nifty, wonderful and fabulous because they are the only known polymers with backbones made entirely from metal atoms. Like polysilanes, polygermanes and polystannanes are being studied for use as electrical conductors.

Polyphosphazenes

Like the polysiloxanes, polyphosphazes are made of alternating atoms, in this case, the chain is made up of alternating phosphorus and nitrogen atoms.

$$\sim\!\!\sim N{=}\underset{\underset{R}{|}}{\underset{O}{|}}\overset{\overset{R}{|}}{\overset{O}{|}}{P}-N{=}\underset{\underset{R}{|}}{\underset{O}{|}}\overset{\overset{R}{|}}{\overset{O}{|}}{P}-N{=}\underset{\underset{R}{|}}{\underset{O}{|}}\overset{\overset{R}{|}}{\overset{O}{|}}{P}-N{=}\underset{\underset{R}{|}}{\underset{O}{|}}\overset{\overset{R}{|}}{\overset{O}{|}}{P}-\sim\!\!\sim$$

An ether substituted polyphosphazene

This backbone is very flexible, like the polysiloxane backbone chain, so polyphosphazenes make good elastomers. They are also very good electrical insulators. Polyphosphazenes are made in two steps:

$$PCl_5 + NH_4Cl \longrightarrow -\!\!\left[N{=}\underset{Cl}{\underset{|}{\overset{Cl}{\overset{|}{P}}}}\right]\!\!-_n \xrightarrow{RONa} -\!\!\left[N{=}\underset{\underset{R}{|}}{\underset{O}{|}}\overset{\overset{R}{|}}{\overset{O}{|}}{P}\right]\!\!-_n$$

Phosphorus pentachloride is reacted with ammonium chloride to get a chlorinated polymer. Then it is treated with an alcohol sodium salt, and that gives us an ether-substituted polyphosphazene.

Sulfur Based Polymers

Linear chain sulfur based polymers are polymeric sulfur and polymeric sulfur nitride.

Polymeric Sulfur

Polymeric Sulfur is prepared by heating rhombic sulphur (S_8)-to 165 – 168°C and then by sudden quenching of the molten mass in ice bath. The final product is washed with CS_2 to remove traces of S_8 because at temperatures below the melting point of sulfur, viz, 114 °C, the polymeric sulfur is unstable and reverts back to rhombic sulfur, S_8. The transition from PS $\rightarrow S_3$ is facilitated by the presence of S_8 in the polymers. The removal of S_8 by washing the polymer with CS_2 slows down the transition. Because of this behavior, polymeric sulfur has limited practical uses.

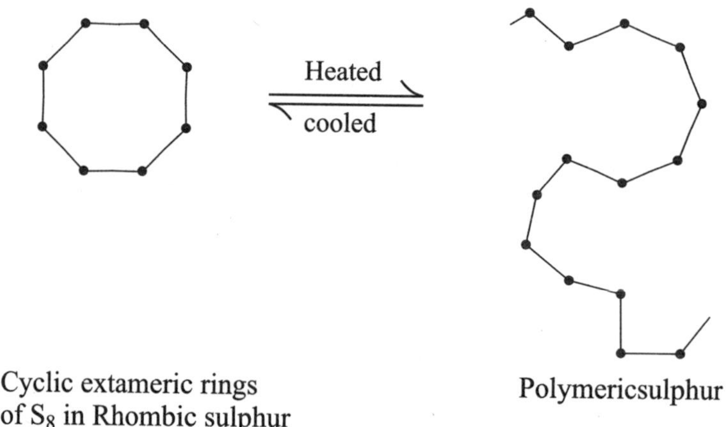

Cyclic extameric rings of S_8 in Rhombic sulphur

Polymericsulphur

Polymeric sulfur nitride

The reaction of S_2Cl_2 with ammonia forms cyclic tetrasulfur tetranitride (A) which is an orange, yellow, crystalline solid which can explode if stored as a solid. (ii) The cyclic tetramer (A) is converted to cyclic dimer (B) by heating to 85 °C in vacuum. The vapors are passed through heated silver wool at 200 to 300 °C. (iii) The solid-state polymerization of the cyclic dimer at 25 °C for 3 days, followed by heating in vacuum at 75 °C for 2 hours leads to the formation of polymeric sulfur nitride (C).

$$\text{(A)} \longrightarrow \text{(B)} \longrightarrow -\!\!\left(S{=}N\right)\!\!-$$

(A) (B) (C)

QUESTIONS

1. Explain the following terms:
   - (*a*) Polymerization
   - (*b*) Degree of polymerization
   - (*c*) Resins
   - (*d*) Monomer
   - (*e*) Homopolymerization
   - (*f*) Heteropolymerization
   - (*g*) Graft polymerization
   - (*h*) Thermoplastic
   - (*i*) Thermosetting
   - (*j*) Silicones
   - (*k*) Elastomer
   - (*l*) Plasticizer
   - (*m*) Antioxidant
   - (*n*) Free radical initiators
   - (*o*) Copolymerization
   - (*p*) Addition polymerization
   - (*q*) Free radical polymerization
   - (*r*) Cationic polymerization
   - (*s*) Anionic polymerization
2. How are polymers classified according to (a) their structure (b) physical properties (c) their response towards heat (d) their mode of formation?
3. What is the difference between addition polymerization and condensation polymerization?
4. What is a Zieglar-Natta catalyst? Give its application.
5. Explain the term graft polymerization. How can it be achieved?
6. Discuss the effect of structure of polymers on its properties.
7. Explain Glass Transition Temperature.
8. Write a note on thermodynamics of free radical polymerization process.
9. What methods are used for addition polymerization? Describe one.
10. What type of polymeric materials are termed as plastics?
11. Explain the difference between thermoplastic and thermosetting resins.
12. For a first order, second order and third order irreversible polycondensation reaction with rate proportional to the concentrations of reactant X and

Y, obtain expressions for the conversion and number average chain length as a function of time for a stoichiometrically equivalent batch.

13. Consider the anionic batch polymerization of styrene in tetrahydrofuran solution, $[M_0] = 1.0$ g-mol/litre, with n-butyllithium initiator. Assuming that $R_i = R_p$, that mixing is perfect and instantaneous, and that the volume of the reaction mass is constant, (a) For an isothermal reaction (constant K_p) obtain an expression relating $\overline{X_n}$ time; (b) The reaction is started with $[I_0] = 1 \times 10^{-3}$ g-mol/litre. At 50 per cent conversion, $[H_2O] = 0.5 \times 10^{-3}$ g-mol/litre. At 100 per cent conversion, what chain lengths will be present in the reaction mass? (c) For case (b), calculate $\overline{X_n}$ at 100 conversions.
14. A polymer has a solubility parameter = 9.95 and $_p = 7$, $_d = 5$ $_h = 5$ and a solubility sphere of radius R = 3. Will a solvent with (= 10.0 and $_p = 8$, $_d =$ 6 and $_h = 0$) dissolve it.
15. The entropy change in forming a solution of polymer is quite small. Explain, why?
16. Poly α-methyl styrene has a higher T_g than polystyrene. Why?
17. Explain the following fact: P.E. and P.S. produced with the stereo-specific catalyst are each fairly rigid, translucent plastic, while 60-40 copolymer of the two product in exactly the same manner is a soft transparent rubber.
18. What must be the temperature to have the viscosity of polyethylene at 300 °F? The activation energy E is 5580 cal/g mol for the polyethylene.
19. By what percentage must M_w of a polymer of commercial interest be changed to reduce viscosity(at low sheer rate) to one-third ?
20. (*a*) Draw flow curves for time-dependent fluids under continuously increasing and then decreasing sheer.

    (*b*) Define the term thixotropic and sheopectic. Give one example each.
21. What is the general relationship observed between T_g and T_m of polymers with

    (*a*) symmetrical repeating units and

    (*b*) unsymmetrical repeating units

    Give two examples of each type of polymer by drawing the structures.
22. Discuss how the crystalline melting point varies with ‘n’ in the “nylon n” series.
23. Describe the role of entanglements in polymer melts rheology.
24. By what percentage must M_w of a commercial polymer be changed to cut in half ?
25. Write down the names of the two common engeenering models to represent flow behavior of polymers quantatively.

26. Define thixotropic fluid and draw flow curves for a thxotropic fluid under continuously increasing and than decreasing shear.
27. Three samples of flexible tubing are subjected to the following tests:
    (*a*) Heated in a bunsen flame. Tubes A and C burn
    (*b*) Soaked overnight in chloroform. Tube A swells, B becomes very stiff and somewhat smaller on drying out, C is unaffected.
    (*c*) Immersed in lead nitrate solution. A turns black while B and C are unaffected. Which of the three is vulcanized rubber, which is poly(ethylene) and which is plasticized poly(vinyl chloride)?
28. You are given a plastic sheet 0.01 inch thick which is stiff and transparent. It burns slowly with a smell of vinegar. It dissolves completely in acetone. Is it poly(ethylene), poly(propylene), poly(vinyl chloride), poly(vinyl acetate), nylon or cellulose diacetate?
29. How would you identify whether a sample is poly(vinyl acetate) or poly(vinyl alcohol) on the basis of IR spectroscopy?
30. Write a short account of the methods used for molding of plastics into articles.
31. How is Teflon obtained? What are its uses?
32. Give formula of polybutylene, polyvinyl acetate and polyisobutylene and their corresponding monomers.
33. What types of resins can be obtained from phenol and formaldehyde? Give the mechanism of their formation.
34. What are epoxy resins? How they are obtained and what are their uses?
35. Explain the terms LDPE and HDPE.
36. Write a note on:
    (*a*) Silicone resins
    (*b*) Polyurethane resins
37. What are rayons? How is it obtained?
38. How is cellulose obtained from cotton and wool?
39. What is regenerated cellulose?
40. Give two examples of synthetic fibers belonging to polyamide and polyesters.
41. What is the difference between nylon 6, nylon 66 and nylon 6 10?
42. How is terylene obtained?
43. Give the structures of rubber and Gutta-Percha.

INDEX

R

S